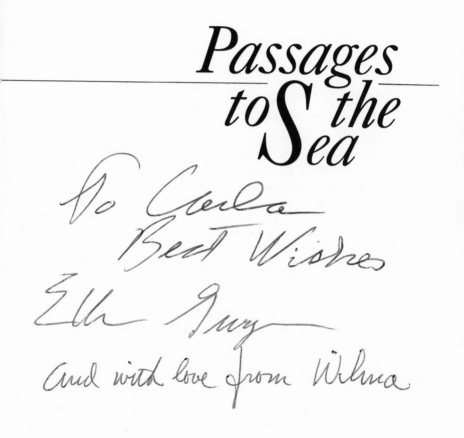

Passages to the Sea

To Carla
Best Wishes

Ella Guy

And with love from Wilma

Correspondence may be directed to:
Ellen Gwynn
2503 Brunswick Road
Charlottesville, Virginia 22903

Jacket illustration by Wilma Bradbeer
Jacket design by Lani Boaz, Papercraft Printing

Printing and type by
Papercraft Printing & Design Company
114 Old Preston Avenue, Charlottesville, Virginia U.S.A.

For I. S. B.

List of Illustrations

Wilma Bradbeer, artist

Notes

At the proper seasons and on similar barrier islands, a visitor will find, if not in the same abundance, many of the sights and sounds portrayed here. Coastland that exists today is not identical to what has recently existed; nor will it remain as it presently is. The narrow sand barriers of North Carolina, unique among coastal islands of the world, advance and retreat in a natural harmony that is not found anywhere else. Holding as they do so many forms of life that utilize both marsh and sea, the pressure of increasing and inflexible human use of the islands makes this life everywhere vulnerable. Let us hope that we will not have to say one day that we should have done more to save it.

A summer hotel built in 1932, the First Colony Inn, stood on the beach opposite Jockey's

Ridge at Nags Head until 1988, several miles north of its present location and much closer to the sea. I thank Betty Clark, a former owner, for making it available to me after it closed in the fall and before it opened in the spring.

It was fortunate for me when I met the late Nellie Pridgeon Myrtle taking one of her early morning walks on the beach, for I was just beginning to learn about the sea. She was both guide and teacher. On a cold September day in 1992, her ashes were scattered as she had wished from a small boat in the vicinity of her beach. Some will remember her for her outspokenness and idiosyncratic ways. I will remember her for keeping a tenacious grasp on essential truths.

My thanks to other residents who have been generous with their time and resources: to Hilda Bayliss and to the North Carolina Marine Resources Center; to Wynne Dough, Curator of the Division of Archives and History of the Outer Banks History Center. Special thanks to Dr. David Lee of the North Carolina State Museum of Natural History for use of his valuable material on the seasonal occurrence of pelagic birds off the coast. I am indebted to all of them and to the various written sources I used in reviewing (and sometimes revising) parts of the book before publication.

But errors are inevitable, and I regret those that despite my efforts to prevent them may have found their way into these pages. The many illustrated and widely available field guides are helpful adjuncts to the accounts of the natural life in the book. Works by the local historian David Stick will satisfy readers who wish to pursue the historical aspect of the area.

My principle geological and geographical references were: 1.) *The Supercontinent Cycle* by R. Nance, T. Worsley and J. Moody, The Scientific American, July, 1988. 2.) *A Sapelo Island Handbook* by B. Kinsey (1982). 3.) *Ribbon of Sand* by D. Alexander and J. Lazell (1992). 4.) *Historical Geography of the North Carolina Outer Banks* by G. Dunbar (1958).

A principle historical source was *A New Voyage to North Carolina Outer Banks* by J. Lawson written in 1709.

Botanical sources included: 1.) *Gray's Manual of Botany* by L. Fernald (1950). 2.) *Vegeta-*

tion of the Outer Banks of North Carolina by C. Brown (1959). 3.) *A Field Guide to the Atlantic Seashore* by K. Gosner (1978). 4.) *Seaside Plants of the Gulf and Atlantic Coasts* by W. Duncan and M. Duncan (1987).

My detailed description of insects was taken from: *The Insect Guide* by R. Swain (1949).

Waves and Beaches by W. Bascom (1980) provided much of the material on these two subjects.

Research papers in the Wilson Bulletin, a professional ornithological journal, supplied some material on the Black-bellied Plover, the Piping Plover, the Red Phalarope, and the Common Loon. An unpublished thesis by William Ihle (1983) and personal communications with researchers who have worked on east coast populations were, in part, the sources of my information on the nesting of Royal Terns. *A Manual for the Identification of the Birds of Minnesota and Neighboring States* by T. Roberts (1949) gave useful information on molts and plumages and details not found in field guides. General references on the status of local birds include: *Birds of the Carolinas* by E. Potter, J. Parnell, and R. Teulings (1980); *Birds and Mammals of the Cape Hatteras Na-*

tional Seashore by J. Parnell, W. Webster, and T. Quay (1991).

The hidden complexity of the commonplace of the region continues to intrigue and surprise; this discovery and the special aura of barrier islands will bring visitors back. Virginia Woolf asked herself in one of her diaries whether life was "very solid, or very shifting." She believed that one single moment might become locked deeply enough within everyday existence to seem eternal, yet, inexplicably, was ephemeral and fleeting. Moments of existence have this beat of intermittency as an ever-present ambience on a barrier island.

Or so it has seemed to me.

Ellen Gwynn
1993

Passages to the Sea

Part One

1

*S*TRUNG ON THE FISHING PIER LESS
than a mile away, pale dots of bulbs stretch into the
sea and glimmer in the distance. Tonight, before the
moon heads for a watch above the marshes, it lays a
path upon the waves from which a vein of light pulls
away and drifts in the shallows, moving with the
waves.

An old hotel made of wood known as the First
Colony Inn is lodged within the dunes. As I sit by a

wide hollow flanked by dunes, not far from the old hotel, plumes of sea oats bend along a gentle slope of sand in my line of sight and angle southward. They bear a fine layer of moisture that the breeze has carried from the water, forming a bright envelope that outlines them with the moon's glow against the white vein of light in the shallows, making each fruit a glistening bead.

As waves slide toward me, I am not aware of the overlying motion of the tide. Yet, the narrow zone between the traces of what the last high tide has left and the reach of lines of foaming water has been gradually widening during the last few hours. This tells me that although a wave may still sweep forward far beyond the place a previous one has traveled, the whole sea is moving in the opposite direction. The moon is nearly full, a gibbous moon.

Eventually will come a high "spring tide," a tide that during storm swells pushes farthest inland. As the earth revolves and produces a centrifugal force that acts on extensive bodies of water, the moon's presence produces the two ebb and the two full tides occurring daily. The nearness of this relatively small satellite gives the moon its preeminence in forming the tides. When the moon is a slim curl, and again just after it grows to a disc, the sun is positioned in line with the earth and moon so that the sun's hold on the sea produces the two most marked high and low tides of a month, the spring tides.

A high spring tide resulting from this horizontal placement of earth, sun, and moon known as syzygy that twice monthly generates the highest and lowest tides can be fueled by a conjunction with perigee, the time each month when the moon lies closest to earth.

Rarely, a spring tide coincides as well with perihelion, that point in earth's orbit where it is closest to the sun, and with the earth's bulging in response to the moon reaching the southernmost point in its orbit. One January, such a rare high tide lifted water eight feet above normal level. Swells rushed through the inlet and up inland creeks. Some miles from where the moon plays on waves tonight, the momentum of this high tide returning to sea at low tide caused water to rise so swiftly in the shallow Sound that artificial dikes were breached from the west, far behind the line of manmade beach dunes. Waves came from the Sound, moving over places where they had not been in recent times, sun and moon finding a strange pathway to the sea.

Abundant signs reveal where, for a time, the sea has withdrawn the play of its tides from certain parts of this region of the coast, despite a slow rise in sea level. Watermarks on bridge pilings or pilings that stand on dry ground are evidence of an ocean that once was higher. The lighthouse of Bodie Island that sends its precisely-timed beacon over thick woods, freshwater flats, and man-made dunes and has its footing planted far from the sea on an inlet that has closed, tells a similar tale. In a place by the present inlet that was the sea's and no longer is, the shoots of the marsh cordgrass, *Spartina alterniflora,* are growing. Quite far inland are signs where waves lapped many hundred of years ago. The slope of a relic dune and hordes of fossil marine shells exposed by excavations dug for bridge pilings indicate recessions of the sea that are very old.

But near the edge of waves along this part of a malleable coast and its wandering islands of sand, the dark line of peat at the base of an eroding dune, or a swath of marsh shells, or stumps of trees whose roots stretch into soil covered by sand, betray the sea's advance. Old photographs taken many years ago of the old hotel and ocean behind it record a beach much wider, and a wooden walkway to it much longer.

Stars can be thick motes of light over the sea when the air is clear, the night behind them a deep black should the moon lie westward, the lack of its light giving the effect of infinite depths behind the panoply of stars. Other nights the sky can be opaque along the beach, the moon over the sea but hidden by mist. The haze is thick enough to mask the moon and conceal the stars, yet it is sufficiently thin to spread a pale light within the low canopy, enabling me to see long distances down the beach.

On certain nights the water can be very dark except for the white lines of the nearest waves. I cannot visually penetrate the cloak over the ocean's expanse; I sense space by some intractable inner eye. The sky is a darkened mirror of the immensity below, sharpening this eye of the imagination, fastening it to the vastness that is the sea's. The rhythm of waves becomes the rhythm of species that have escaped contingencies, the lives created and re-created to forms of increasing complexity since the coming to ancient Cambrian seas of a tiny swimmer with a bone-like rod in its back.

I am tempted to believe that all these ancestral lives are inseparable from what I am, and some lodestone of consciousness resonates to an unsuspected mode of existence.

One bird of the beach is part of its darkness. The large luminous eye of a Black-bellied Plover watches me as if it guards some inscrutable secret of the sea I would never perceive. The plover is feeding where a sheet of foam has just washed, a few brief and hurried steps, a pause, a quick probe—a way of foraging usual for it. Poised, alert, careful, attentive to every nuance around it, forever a step beyond my close approach, the Black-bellied Plover embodies in its passage to the sea all that is cryptic, and the process that has probably created and compressed this sea in cycles has whispered something into the night of the puzzle of contingencies.

I know of no other bird that will call frequently

in the night as if to ascertain by the clear mellow whistle that the darkness is its own, and no one else's. The long-legged shorebird is solitary in habit. It calls for what? For whom? Another plover may answer, and there is something touching in the intensity of voices.

The plover whose liquid note haunts the darkness is found during all seasons. It is seen in a gray garb on the coast for a large proportion of this time. In summer the bolder contrast between the back and the underparts is most visible. By the time birds of this species that have nested earliest on the tundra return to the beaches near the old hotel, those that have waited until very late to go north may still be here.

The brighter plumage of this dove-sized compact plover with a bill like a dove's is worn for the Arctic nesting grounds. Back feathers of gray, marked with brownish black, contrast sharply with the unbroken wash of black on the breast that reaches from chin to legs. By August a reversal to the gray dress has begun. The lattice on the back dulls to shades of gray, the black underparts turn a mottled black, then whitish gray.

A Black-bellied Plover is usually attached here in the vicinity of the old hotel, and forges a wary link when I am on the beach. The large eye so uncharacteristic of most other shorebirds carefully will track my movements. The Black-bellied Plover ordinarily forages on wet sand, but it will feed on clumps of vegetation higher up on the beach, or in the deep narrow run of a tire track, or near the base of the dunes, sometimes finding grasshoppers or crickets in the dune grass.

Other shorebirds are not as apt to utilize these niches as sources of food, preferring to feed closer to the waves. Most of the time they travel and forage in small groups on the beach. Their feeding is more sustained and their absorption with the task more complete. They are also more tolerant of being approached. In some cases the smaller shorebirds are unusually tame and have a preference for grouping with species other than their own. Even on its arctic nesting grounds the Black-bellied Plover is not gregarious. Yet one species, the Knot, chooses deliberately to nest in close proximity with this plover, despite the fact it is not made welcome.

Another kind of plover is very similar to the Black-bellied, both in its Arctic nesting plumage and the subtler one worn in fall and winter. During migration it is not impossible that the Golden Plover and the Black-bellied will be glimpsed together on wave-washed beaches, and on mudflats in the marshes.

When in the nesting plumage, the Golden Plover has the black underparts extending to the tail, and is the true "black-bellied" plover. The feathers of its back are browner and flecked with golden spots. Juvenile Black-bellied Plovers briefly wear in fall some of these yellow markings of the Golden Plover; the golden spots persist on the wing coverts and rump after the fall molt of young birds into their first winter plumage.

Before Black-bellied juveniles leave the waters of this coast and begin the long flight that will take them far north, the large feathers of the wing are replaced. Males in their first acquisition of the bolder nesting plumage lose more of the pattern of the winter plumage. Generally, the sexes are alike. Although the Black-bellied Plover is the largest of the plovers,

11

stress in an individual bird reduces its size. A male often incubates the eggs and raises the young alone. A female lays four eggs in as many days. This is 70 percent of her weight; within a single season she can lay in a second nest. Two Black-bellied Plovers found feeding in the same vicinity on a southern beach can vary in size enough to appear as separate species. This is true of individuals of other shorebirds that for one reason or another have not been able to store lipids or fat.

The plover that has been watching me while I sit quietly by the dune hollow moves from the wave-washed rim of the sea, but does not come close. The dipping whistle it gives as it flies off is unmistakable; no sandpiper here calls this way.

When I send the beam of my flashlight in its direction, my gaze goes to one particular place before the plover is lost over the sea.

I have only the briefest look at the bird as it crosses the bright track of light. For an instant, a crescent of black wedged below the place where the wing meets the body shows clear and sharp against the wing's white lining and against the light breast.

The clusters of black feathers are vivid signals during the day if a Black-bellied Plover is in flight. Should I happen to look up to find one high overhead as it crosses to the sea from the west, or to the west from the sea, these slices of darkness that are usually buried under folded wings are precise and unfailing markers. Only while the bird is in the winter plumage are they so powerful a magnet as soon as the wings are opened. When the underparts turn black in the spring, the prominence that a light breast has given these dark clumps of axillar feathers is lost, but they are always with the bird.

The Black-bellied Plover was once called the Gray Plover, named for the duller plumage it wears for the greater part of the year. This plumage blends the Black-bellied and Golden Plover together more closely.

Only one of the plovers habitually frequents both sea and marsh. The plaintive whistle that takes hold of the night belongs only to one. Only one keeps the darkness of the moonless sea beneath its wings.

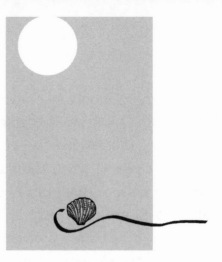

2

*T*HE TIDES THAT EXPAND AND CONTRACT the sea may murmur of a subtler, richer pulse tripping and subsiding, a slow primordial tempo making the Atlantic a transient sea.

If other supercontinents like the most recent landmass known as Pangaea have existed in the past, the Pacific is the immoveable and outer ocean of the earth, the Atlantic the inner one that appears and vanishes.

Some geologists believe that oceans similar to the Atlantic have swelled among the dispersing fragments of a succession of supercontinents like Pangaea and have closed when the fragments rejoined. This massive pulse is thought to beat in periodic cycles of about 500 million years.

The new theory proposes that while the Pacific has remained in place, an Atlantic-type interior ocean has opened at least several times in the history of the earth. The pressure of trapped heat from a molten layer below the earth's crust forces a large global landmass to dome and then to become cracked where pressure is greatest. Uneven convection of heat eventually splits the landmass and allows the release of magma.

Heat continues to escape through the floor of an interior sea that begins to form among the fragments. Gradually the sea widens and deepens, and the fragments disperse on crustal plates.

If a more precise timetable for the theory is correct based on both the geological record and current data, the floor of the Atlantic has not yet begun the movement of plates that will rejoin the fragments of Pangaea. The earth is midway in the beat of a great heart. Where heat is escaping, the sea floor continues to grow heavy and dense with cooled magma; its growing weight will cause the edges of crustal plates to slide beneath adjoining lighter plates to which continents are fastened, drawing fragments together again.

The force of rising heat building beneath the landmass of Pangaea began the present cycle. Rifts appeared within 80 million years. An additional 40 million years of pressure split the landmass, and the continents moved away from one another as extruded magma wedged apart the ocean floor.

In accordance with the closer estimate proposed for a cycle, continents reach a distance in 160 million years beyond which they separate no farther. Another 160 million years must pass before the reprocessing of the edges of subducted ocean plates to magma once more constructs a new supercontinent.

The regular emergence of large mountain ranges have helped to define supercontinent cycles. In the case of Pangaea, the rejoining of two fragments of the previous supercontinent created a range rivaling in height and grandeur the present day Alps. The Appalachians were born and on Pangaea grew older. Nearby lowlands slowly accumulated sediment that was worn from the rock of the massive mountain peaks by wind, rain, and severe fluctuations in temperature.

Once the breakup of Pangaea into the present configuration of continents occurred, the lowlands containing the sediments eroded from the Appalachians were uplifted. They became the region of North America known as the Piedmont that lies near the east coast. The aging of the ancient range of mountains is part of the story of North Carolina's barrier islands. Exposed to erosion by inland streams and rivers near the coast, some of the eroded material forming the Piedmont was carried to the Atlantic and began to accumulate on the continental shelf.

Elevations of sea level in warm interglacial times, when ice fields returned their frozen catches of water to the Atlantic, and a fall in level, when colder temperatures again locked water in glaciers, produced greater amounts of sediment that became the matrix for the earliest barrier islands and the present ones along North Carolina's coast. A malleable plain of alluvial sand was molded into attenuated elevations strung together in the sea by a process only recently understood. Islands of sand have oscillated here for the past 100,000 years whenever a shallow continental shelf was present. It was believed that such islands were left by a retreating sea. A current view proposes that a barrier island is dismantled when sea level falls and is created again when sea level rises. An island prevents the full play of high energy ocean waves westward. On its lee side where wave energy is low, fine clays are able to silt out from rivers and accumulate enough depth to hold vegetation, which, in turn, binds greater depositions of clay.

In general outline, the present coast of North Carolina has been in place about a thousand years. However, maps drawn as early as the 16th century indicate that natural passageways have intermittently joined the sea on the east and the marshes and Sounds on the west. The sea does not war with land in the way that it must on the more forbidding rocky coasts of northern states. Long, low dunes are regularly crossed along the island chain. Bodie and Pea Islands, which are currently separated by Oregon Inlet, have

19

had a more or less continuous adjustment in the past several hundred years to the combined pressure of tidal and storm seas seeking a way inland and of inland waters seeking outlets.

In the 16th century an early version of Oregon Inlet opened and in the following century closed. South of Oregon Inlet, "New" Inlet has twice opened and twice closed since the 18th century. To the north "Roanoke" Inlet once joined the world of sea and that of the Sound and marsh. Thirty-five years after Roanoke Inlet closed in 1811, Oregon Inlet opened for the second time, and with the usual vacillation that channels have shown along this particular stretch of North Carolina's barrier islands, this inlet is now filling with sand and shifting south.

With skeins of estuarine marsh and Sound on one side and the Atlantic on the other, a reciprocity of land and sea exists that is unique among coasts of the world. Unlikely travelers with a magical adeptness at bridging two very different spheres make a harmony of opposites seem so easily theirs, dividing themselves between a wave-filled sea on the east and marsh and Sound on the west.

Their need for both an open world without boundaries and a more solid and contained one is an easy, porous balancing. The sounds of shorebirds often come as I walk at night, and I am able to distinguish their shadowy forms from large ghost crabs in the churn of water gliding over sand; sometimes as many as half a dozen birds, usually just a few dark shapes piping like ghosts. The old building behind me is heavy in the darkness, the scent of sea thick on the air. I listen to the brushing of wind steadily erratic that lifts water, filling it with an ancestral rhythm. And, if, on the other side of the hotel where leggy Pittasporum shrubs grow, the music of insects is silvery and bell-like in every cranny of a summer night, hanging in the air yellowed by the ceiling lights of the veranda, waves are the sea's silvered shakings.

A moment comes, bittersweet like these voices of summer, in which the deep and lasting pull toward the illimitable completes and makes whole some inner dichotomy; the rhythm of impermanence exists as a state of mind, an interior symmetry caught and distilled in the sound of waves.

Atlantic Bottlenosed Dolphins can pass at any time of day in summer, feeding as they travel south and back again, but I see them regularly in the early morning hours. At that time they often come in very close to shore in front of the old hotel, circling along the innerside of the small sandbar that lies only a short distance out. They are mammals, and unlike the sea turtles that must return to land for one night of egg laying, dolphins need make no landfall, birthing and nurturing their young in the sea.

The continental shelf edge of 100 fathoms, followed for a short distance by the Gulf Stream, and the deeper water adjacent to it by which the Gulf Stream flows, form the center of a strange "city" where these dolphins roam with other kinds of dolphins that are not seen from shore. At various seasons of the year, whales of different species migrate there.

If the sun is high and the sky clear, the watery film on the fins of Bottlenosed Dolphins catches light and splinters sun into diamond facets sparkling in the water. Late in the day, the dolphins pass in dark lines, alternately diving and rising in brief rolls as they forage. The nearby waves belong to them. Beyond these strings of dolphins and even beyond the farthermost lines of waves, the offshore waters of North Carolina also hold a diverse aerial life whose existence a visitor does not suspect. Boreal seabirds make this city of the sea the southern limit of their range. Tropical and subtropical seabirds make it the northern limit. At Cape Hatteras, the continental shelf edge and the western side of the Gulf Stream come relatively close to land, producing a fertile feeding ground about 35 miles from shore. Near Oregon Inlet, the Gulf Stream bends from its northward course that has taken it parallel to land and swings eastward. Above the Gulf Stream's eastward swing is an area of "green water," a deep zone as fertile as the sunken edge of the continent or the shoreward edge of the Gulf Stream's blue water. It completes the unusually rich linear stretch along North Carolina's coast.

In other coastal states, the continental shelf extends farther eastward. The shelf areas of those states in summer have been found to be relatively barren of a diversity of seabirds other than gulls. It was assumed that, mile for mile, the offshore waters of North Carolina in summer were as sparsely populated.

Had it not been for the work of one researcher, the long narrow city in the sea and many of its inhabitants would be virtually unknown; for beyond the sight of land, this unusually fertile stretch lures birds that make extensive, temporary flights in summer from the islands where they nest.

Sporadic records do exist documenting birds from the tropics and subtropics that the revolving winds of hurricanes have accidentally trapped and transported to the east coast. Birds are rerouted during migrations by strong and steady blows from the east. The subadults of the Greater Shearwater, sea birds from thousands of miles away near the Tropic of Capricorn, become stranded in large numbers on the coast, if they have become weakened during their flight north by having been delayed by lack of wind in the doldrum latitudes.

The records tell little of the natural rhythms just beginning or just ending in local coastal waters, rhythms embedded in the lives of ocean-going birds known as pelagics. As far as it is known, only at North Carolina does an extraordinary offshore environment created by complex movements of the sea attract and sustain a community of various summer birds each year.

Although not encountered as frequently as other pelagics, the bird that has been called the "Wide-awake," the Sooty Tern, inhabits the city in summer. This bird has been given the name Wide-awake because of activity uninterrupted by sleep during the night at nesting sites on islands off Florida's tip or on islands in the Bahamas. It is possible that parent birds fly the distance to North Carolina's enriched waters, feed, and fly back. But perhaps the name Wide-awake belongs more to the young of this bird. When they are able, they journey to these feeding grounds that are more fertile than the waters around the nesting sites. Unlike the adults that come here, the young Sooty Tern is seven years at sea, and wakeful for these years. It will not land on the sea. It feeds by swooping close to the water and snatching small squid and fish flushed to the surface by the activity of predators. If it rests, it must find flotsam to support its weight. The plumage of young and adult Sooty Terns apparently do not have the sealants possessed by other pelagics.

This city of the sea seems an unforgiving place. But surrounded as it is by waves, it is not without its sanctuaries. A tern that like the Sooty Tern neither swims nor dives, the Bridled Tern of the Caribbean, finds them. The young of a Bridled Tern family were seen by the researcher as they sat on a drifting board calling to their parents to be fed: and this tern is known to perch on flotsam. Both the Sooty and the Bridled Tern have the thin black "bridle" that gives the Bridled Tern its name. This line extends from the corner of the mouth to the eye and meets a black headcap. Only the Sooty Tern has the rich stove-black mantle on the wings and back. Until recently, both these handsome black and white terns were considered accidental, their presence off North Carolina's coast attributed to the winds of tropical storms.

Regular seasonal sightings of Bridled Terns by the researcher indicate nested birds and foraging birds with young are neither rare nor haphazard, and the journeys these terns make at various seasons of the year are a normal extension of the nesting cycle.

The same kind of journey is true for the Bermuda Petrel. This is a rare petrel numbering less than 50 pairs and now protected on the last nesting site of what had previously been a wide range of them that included islands in the Caribbean. It is found on one island of the Bermudas, the east coast's only oceanic islands, which lie at the same latitude as South Carolina, a distance about as far as that which the Sooty and the Bridled Tern fly.

Superficially, the seabirds called petrels are gull-like and not tern-like, with plumages that are generally dark on the back and lighter below. A pair of conspicuously raised tubules on the upper bill function to expel salt, and, in the case of some species, to eject noxious substances that deter predators. For other petrels, the pair of tubules refines the sense of smell, a sense usually undeveloped in birds.

A petrel that some years ago was thought to be extinct, the Black-capped Petrel from Cuba and Haiti, comes to the city in the sea off North Carolina's coast. On one day, thousands of these birds were sighted here along the 500 fathom contour. The long narrow wings of the Black-capped Petrel are able to execute swift upward rolls and plunging drops and long tilted glides, generating on occasion speeds in excess of 70 miles an hour. The only known place where concentrations of this petrel have been discovered so far is off the coast of North Carolina, suggesting that it, too, makes flight from the island it nests upon to find food that the waters around the island cannot supply.

The Audubon's Shearwater, also a tube-nosed bird, journeys from the Caribbean. Thousands of these birds have been seen in late August. The long tail and short wings of this small shearwater make it appear more like a gull than most other tube-nosed pelagics. It favors the inner edge of the Gulf Stream and areas of water 50 to 500 fathoms deep and, in particular, current edges with floating mats of Sargassum.

To the city in the sea comes the White-tailed Tropicbird in summer. It frequents the warm shelf waters south of Hatteras and those near the 100 fathom contour of the Gulf Stream north of Hatteras. But for the elaborate dress that it wears while wandering the waves, this tropicbird is more tern-like in the shape of its bill and body, and, like the large tern native to the coast, the Royal Tern, catches fish by diving into the water from the air. The exact nesting site from which the White-tailed Tropicbird comes is not yet known (or is this known for most seabirds); it flies from either the Bermudas or from islands in the Caribbean. A narrow, black brush stroke accents the white of its wings and aligns visually in flight with the narrow pair of white feathers in the central part of its tail. The bird is said to sit upon the sea with these two 16-inch tail streamers held high, preventing them from contacting the water.

From the Caribbean comes the Masked Booby. It is a smaller relative of the Northern Gannet that is commonly seen from the shore in winter. The adult is white-bodied, with a deep black border running along the back edge of its white wings and along the edges of its tail. A distinct although narrow dark mask circles the short feathers of its face near the bill.

From far below the latitudes of the Caribbean, the Greater Shearwater comes after nesting on islands such as those near Tristan da Cunha, which lie south of the Tropic of Capricorn. In the years when the doldrum latitudes stall the subadult birds and no winds blow, they reach the coast of North Carolina starved and weakened. Normally, wind will "scale" a healthy Greater Shearwater over the waves for the time it spends in North Carolina's waters. Air deflected from walls of water and pushed against a shearwater's long narrow wings, which it holds very stiff and slightly bowed, will in theory fly a bird all day without the wings beating; when it begins to slow down, it gains altitude by turning sharply into the wind for another close slide.

The inhabitants of the city in the sea come not only from South Atlantic islands as far away as Tristan da Cunha. Having made its long flight from the sub-Antarctic, the little Wilson's Storm Petrel is found in local offshore waters in summer and is most commonly seen within a few miles of the 100 fathom contour. Its nostrils are distinctively fused in one tubule. Chum attracts this petrel and it will follow boats. It has the habit, too, of pushing off from the water with a characteristic patter of its yellow-webbed feet, and, using the same motions, of fluttering over the water.

In summer, the rich feeding ground off North Carolina's coast draws pelagic birds from across the entire width of sea. From some of the islands off the west coast of Africa, possibly those of St. Helena or Ascension Island or the Azores, flies the Band-rumped Storm Petrel. Its plumage but not its behavior resembles that of the Wilson's Storm Petrel. Until recently, its appearance on this side of the Atlantic was considered accidental, the sporadic records of it here, in some cases, thought to be the result of the bird becoming sealed in the eye of hurricanes and being forced to move in tempo with these huge wind traps. Although Band-rumped Storm Petrels are a warm water species, a few of the sightings made by the researcher occurred over the deep "green water" lying east of Oregon Inlet. In summer, the presence of this petrel went undetected until collections were made and identifications at sea verified.

From sites across the Atlantic's width, such as the Azores and Canary Islands, comes a shearwater resembling the Greater Shearwater, but it is without the white collar possessed by the slightly larger Greater Shearwater. The Cory's Shearwater is also a regular and numerous inhabitant during summer.

North Carolina's offshore waters are the best studied region of the Atlantic in this hemisphere for both mammals and seabirds. Nested and postnested birds with a cycle of long sea flights have adjusted to contingency in the folds of the waves, the floating weed mats and flotsam, the edges of fluid roads, and the pumping wells of deep enriched water. Birds feed in diverse ways on the life that comes in abundance. They skim off the natural slicks and "waxes" emitted into the sea by fish and large mammals; they hunt small fish, squid, jellyfish, and other fare.

The migration routes of millions of song birds take them regularly out to sea. Few reprieves exist offshore for the land birds of garden and woodland, birds that are better known to visitors of the Outer Banks. One spring, after an east wind, I found in the night's windrows the single feather from a Cardinal. Signs of any voyage ended in the sea are not normally encountered. A downed bird will float for a time because of the air in its hollow bones, but the temporary buoyancy is lost when the plumage begins to absorb water. Waves that close over the graves of exhausted land birds will hold aloft pelagics that swim or rest there; the smooth walls of waves will scale the long, bowed wings of shearwaters for miles and miles of sea.

In this place of passages that has been thought to be relatively empty and visited only by migrants from land or by pelagics whose routes cross it, the discovery of the seasonal assemblies of so many different birds is remarkable enough. A further thread of the story is even more astonishing. Tiny riders are carried there by seabirds. Some of these spiderlike mites living on the body of a pelagic exist on the bird's tissue, some on the desiccated material from the skin and feathers. I know of no journey made by so small a voyager nor do I know of any stranger than that of the mite that feeds on the fare of the sea when brought to the world of flux.

The little rider consumes microscopic single-celled plants called diatoms that secrete a shell of silica and that fill a drop of seawater by the thousands. While scaling waves, a pelagic such as the shearwater accidentally collects water drops at the ends of its wings. To reach the drops filled with diatoms, the mite has learned to migrate down the feather shafts to the feast of plants that awaits it on the feather tips.

The researcher believes that the epic flight some seabirds are forced to make from less fertile waters near their nesting sites to the more productive stretch off North Carolina has its answer in the breakup of Pangaea and in the diverse routes taken by the fragments of earth's recent supercontinent. It is possible that long ago the stage was being set for the summer visits of pelagics to the city in the sea. When the landmass of Pangaea fractured from convective forces that acted unevenly, India headed northward for a meeting with Asia; North America initially was still part of Europe. Africa began easing from the eastern side of South America, and Madagascar from the eastern side of Africa. Australia was yet to unnotch itself and to move northward from the Antarctic. Within 135 million years, the separation of continents allowed room among them for basins to form and to widen enough to hold seas.

As North America moved westward from Europe and Africa moved eastward from South America, the great interior sea, the Atlantic, was formed. Continental movement may have shifted the nesting sites of ancestral stocks of some seabirds away from rich coastal feeding grounds. In that event, the birds that did not seek new nesting sites were without fertile waters close by after the continents came to lie in their present position. Loss of the good foraging waters that the drifting continents had left behind then forced regular flights by these birds to new ones.

Whatever the reason responsible for the seasonal flights to North Carolina's offshore waters, the flights are commonplace here, and as is so often the case with the commonplace, there is much that intrigues, much more to be learned of the birds that make them. On summer days at high tide, I have crossed the zone of large waves coming over the bar and spilling shoreward, and bobbed just beyond them where the sea was calmer. Beneath my back the waves of the incoming swell lifted me as I floated. I could look shoreward and see the familiar tiers of the roof of the old hotel, the long, white, veranda railings and, then, with a turn of my head, the open sea. The waves that lap against the sky, curling and shrinking on the horizon, say nothing of a surprising gathering of life in a place that has kept its secrets so well.

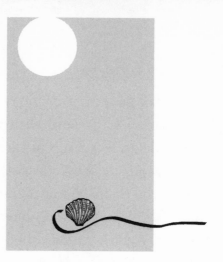

3

*F*OR AS LONG AS SEAS HAVE BEEN ON EARTH, geologists tell us, there have been waves. Before there was a wind, disturbances in the rock basins holding the first seas produced primordial waves whose anatomy, had anyone been present to witness the dynamism, was a silent search for restoration upon empty shores. Wind waves formed on the early earth with the beginning of motion in the new atmosphere. Evaporation of ocean water and the condensation of this water generated cyclic rains. An atmosphere began changing the land with winds that were strong and wide and capable of carrying large amounts of water vapor as they traveled. Fast-flowing streams eroded the rocks of continents. Gravity brought this material from upland regions to the seas. Heavy material stayed near shore; the light drifted farther out. The edges of the continents were worn down by waves driven by wind.

Silent waves exist that move under the sea on the interface of fluid layers differing in density, betrayed only by "dead" water or by a characteristic reflectivity. The smallest known waves are voiceless and just beyond me if I stand at the edge of the sea. On the slopes of ordinary gravity waves or on a glassy sea brushed by a light wind, the elastic quality of the water's surface generates tiny wavelets. The distance between two crests, the wave length, is but a fraction of an inch, and the time it takes a crest to travel the distance is a second at most.

Waves with the longest known wave lengths reach halfway around the circumference of the earth, and they, too, are voiceless. They are the tides of the sea. The time it takes for such waves to travel the distance between two crests and to widen and narrow the coastline is 12 hours and 25 minutes.

If the wind is strong enough, waves will crest and spill far from shore in a thundering upheaval that overlays a tide and forces one wave over another. On days when waves are continuously breaking from shore to horizon without rhythm, so that no stretch of water is even, what seems a random building of sea when the eye tries to choose but one crest or one trough to focus upon and follow before either becomes lost in another rise and dip, could the eye decipher it, is a variety of individual wave histories superimposed on each other.

One series tells the tale of a storm that was distant, already over; a second, of a nearer storm, from a different direction and still raging; yet a third series, of a storm just gathering. Each story is as varied as is anything of the sea, all the histories legible if it were possible to pry one from the rest. Without warning, the apparent disorder may generate the great lone marauder that assimilates in one towering mass the waves from widely separated storms and suddenly rises to a hundred feet and as suddenly flattens again on open sea.

Ordinary surface waves are best observed when the tide neither ebbs nor flows in a sea without wind. On a series of waves generated by a passing boat, a raft of gulls floating nearby will rise high on the crests, will dip deep into the troughs without ever moving inland, what each particle of water the birds float upon does in its circular gravitational orbit. The waves only temporarily dislocate a particle and the raft. When the waves reach a shoaling bottom, the simple pattern of gravity acting to restore water to its original level and to fill up the trough ahead causes a wave to change its form. The crest steepens, the distance between two crests becomes shorter. This, in turn, increases wave height as the top of the wave begins to move more rapidly than the water below that is slowed by shallows. Finally, at a depth of a little over twice the height of the steepening crest, the wave collapses, tumbling forward.

In the surf zone, white and bubbling with entrapped air forced into the water by waves breaking, a wave can re-form again if there is a channel near shore. Waves collapsing over an inshore bar and rising again over deeper water hold a paradox. A crest there actually speeds forward toward the finish of its journey. No longer oscillating in a circular motion, without a trough and nearly soundless, the wave ends as a foam line that leaves a narrow undulation of raised grains of sand on shore, the brief, tangible mark of ceaseless repair and the script of an ancient language spoken to a different world.

Improbable as it is that any creature could tune itself to this rapid flat sigh of the sea, a small bird of the coast, the Sanderling, has become adept at feeding by foam lines. The little reflection of a wave's final whisper has just come to the sea's edge. It may have already probed today near the marsh for the succulent life beneath the mud, without hurrying, without need of protecting its feathers, or is it that the practice of a specialized technique of feeding near waves causes a Sanderling to appear to escape them?

I do not have long to wait before the Sanderling is joined by others. Like a net thrown over the water but not dropping, shivering and shuddering and retrieved, a flock near shore strikes out on an oblique path and turns back, veers away and reverses course again before landing. Immediately the birds run along smooth sand in a cluster that expands when a wave slides forward and closes when it retreats. One of the birds succeeds in taking a quick bath in a narrow tongue of foam. The effort is very brief, and the bird collects itself quickly and goes on with the others.

When Sanderlings are not characteristically galvanized to the rhythm of waves, they are one of the difficult species of small shorebirds to identify. Adult birds in a flock can wear the gray winter plumage or the darker rusty-brown nesting plumage; young birds can also be light or dark. A number of small brownish sandpiper species unfortunately will readily attempt to feed the way Sanderlings do. It takes a practiced ear to know the calls of these sandpipers so similar in appearance and to use the slightly different call of the Sanderling to make an identification. But there is one characteristic that distinguishes both young and adult Sanderlings from the confusing shorebirds that share their general size and color. A wide unbroken stripe of white begins at the back of a wing, widens at the bend, and continues toward the tip. The stripes are visible as soon as a Sanderling flies.

Sanderlings are unique among sandpipers in having a streamlined foot bearing only three toes. Perhaps it is the added agility a foot that has lost the fourth toe gives and a keen sense of the sharpening and softening of escaping air in a sheet of foam that combine in an experienced bird to perfect the ability to skitter close to waves. Rolling wheels instead of legs seem to carry the adult Sanderling ahead of foaming up-sweeps without it ever being touched by water, wheels that lock and, in a split second, revolve in the direction of down-sweeps. There is very little time to work the patch of sand from which a wave has slipped before a second wave comes.

I think it is less a matter of avoiding water than it is the briefness of opportunity to mine the space, for in it Mole Crabs are more successfully seized by the slender bill. The group of Sanderlings that have chosen this part of the beach on which to forage are finding Mole Crabs. The crabs live in the mercurial white strip of foam-covered sand that is always in motion, extending their feathery antennae into the push of percolating water to collect organic matter; only for a few seconds are the crabs visible. They then burrow swiftly out of sight in the bare, water-loosened sand after a wave passes. A patch of washed beach filled with little Mole Crabs bubbles and shifts under a thin glaze of froth and shines smooth as silk again; if my eye is vigilant and my hand quick, in a clot of sand I may hold a score of these animals that live at edge of the sea.

The long narrow bill of a Sanderling shaped to plunge rapidly in and out of sand is much like the bill of other sandpipers that also feed on Mole Crabs and other small forms of shore life. If look-alike species of shorebirds are feeding in a Sanderling flock, they give themselves away when the flock flies up and some of the airborne birds are without a clear wing stripe. Sandpipers other than Sanderlings usually follow waves only haphazardly and move parallel to shore. The feeding flock is composed entirely of Sanderlings. Not all are adept at their highly specialized mode of feeding. I suspect that it is a young bird that takes a Mole Crab much too large to swallow and thereby attracts a young pirating gull, which pursues the Sanderling for long distances up and down the beach in an erratic chase that doesn't cease until the Mole Crab is dropped.

The center part of a juvenile's tail appears much paler in contrast to the same area in the tail of an adult. One of the Sanderlings, too riveted at its rummaging, is forced to fly to elude the surge of water that comes streaming over the sand, but I do not get a good glimpse of this bird.

Some of the young gulls that have been watching attempt to mimic what they see. With guarded steps they walk into a sheet of water and stab at the sand, retreat a little and try it again with steps markedly firmer, probing the sand when a wave recedes and following the wave down.

The flock of Sanderlings draws a group of grackles to try the method and they alight among the flock, walk into the water and out again, their keeled tails caught by wind, turning them in directions they do not wish to go. Unable to maneuver because of the hindrance, the grackles keep the experiment brief, but stay with the Sanderlings for a while in shallow water, recognizing that there is food here. Ungainly hops rescue a bird from deeper water as it searches for the crabs the Sanderlings are feeding upon. One of the grackles secures a Mole Crab so large it cannot be swallowed. Without it being bothered by gulls, the grackle carries the crab to a wood railing and begins to tear it apart with its strong feet until the soft interior is exposed. The meal is not finished before the remaining parts are carefully pried into smaller pieces as one foot holds them, and the claw of the toe of the other separates them.

After a while the Sanderlings leave the edge of the waves to the opportunistic gulls and grackles and rest in a group on the upper beach, one or two of them nodding and drowsy. Some fall asleep and are suddenly collected, having been awakened by the restlessness of companions responding to the intrusion of a gull. The birds take their heads from under their wings without lowering the foot they have tucked into their bodies. Loath to use both of the legs that have recently been revved and kept to a feverish pitch, a bird woken from a nap and bothered enough to move keeps the weightless leg folded as it hops on but one.

The flock will not rest for long; night does not suspend the marvelous precision of these slender, delicate legs. And the Sound and marsh, the refuge from everything that is sea, might beckon the birds before tomorrow.

Behind the high dune of Jockey's Ridge, spits of sand lie in Roanoke Sound. I have seen Sanderlings come from the sea and seem not to be able to stop the ticking in their legs, as if "pre-wound" by waves and still following the rhythm to nowhere in particular on a spit. Eventually some of the birds rested or bathed leisurely in the waveless water edging the spit, ruffling and settling their feathers. The cleansing ritual allowed feathers to reseal, and the grooming necessary for flight and for a waterproof covering that resists saturation took place where the air and water were still.

I wonder if there was some need, other than hunger, to follow waves, some groove in their brain that re-activated the memory of moving their slender legs, and that need returned the Sanderlings to the sea. Young birds learn to master the timing of waves, to flow more smoothly with the swift

47

climbing and falling back of sweeps of foam. Each time they leave the motionless world to the west, they follow the whisper of waves more surely, cohering by their rapidly moving feet to this ancient language.

Part Two

4

WHERE SUBMERGED OFFSHORE LAND is neither rugged and craggy nor markedly hilly, but is sloped gradually toward the coastline, beaches composed mostly of quartz and feldspar stretch along the Atlantic as an extensive system of barrier islands. Under these conditions of below water terrain and because of their very number, the minute grains forming islands have produced over the long term a generally stable configuration of marsh and sea, although

53

individual islands shift and separate.

Lying west of the summer hotel, the huge rise of sand with scattered valleys that is known as Jockey's Ridge is an unusual landmark of the coast, and nowhere is the presence of such an accumulation of sand more visible. Unstabilized by the planting of grass and for this reason unlike the large dune upon which the Wright Memorial stands some miles north, the ridge is a "live" dune, a rolling land mountain belonging more to the marsh than to sea.

A series of aerial photographs begun in 1945 shows the changing momentum that shifts the ridge along the strip of land between Roanoke Sound and the Atlantic. Sand is moved back and forth seasonally along the strip by the see-saw path of prevailing winds coming in summer from the southwest and in winter from the northeast. The ridge built itself from extensive sandy shoals that strong ocean currents, severe storms, and hurricanes deposited on the beach. Collected by wind and blown into mounds that have stood higher than 140 feet, the dune has remained where it was formed because of a lack of wind constant enough to disperse it.

About half a mile away from Jockey's Ridge was a prominent dune and seven adjacent hills that have disappeared into the Sound. To that region of live dunes planters came from nearby counties and were perhaps the first who summered there. By the middle of the season fields had been sowed, and a gathering of families that otherwise would have been isolated by the long distance between plantations vacationed during the hot weather. The first "colony" occupied the soundside south of Jockey's Ridge for a distance of about a mile. The sea was far away, but visits were made twice daily by means of a mule-drawn wagon with a cloth top. Life in the resort community thrived and was only temporarily interrupted by the Civil War. Insulated as the community was on the shores of the Sound, it was able for a time to preserve a mode of life that elsewhere had begun to change.

As much of a reason for the colony's move to the sea as were the construction of roads and a bridge, the growing use of automobiles, and the ever-mysterious pull of the sea, was the capriciousness of the hills that sent their waves of sand over entire houses, engulfing them. The remains of buried cottages lie in the present

Sound, as well as the remains of those whose pilings were sheared by ice formed near shore in winter if temperatures stayed low and that then broke up in destructive blocks during a thaw.

Each year Jockey's Ridge seems to be diminishing in height. The equalizing forces that made the placement of the dune more or less stable have not changed as much as the sandy coastline that has yearly renewed the ridge. The loss of beach to roads, to buildings and access driveways, and to vegetation that has increased because of a more protected environment, prevents the necessary transfer of sand to the ridge. Due to erosive seas, beaches themselves are no longer the liberal suppliers of sand they once were. At the dune's south end, dark patches of large grains from which the pale and small ones have been skimmed by wind, and the changed contours of peaks, quite different after a night of a strong northeast wind, show the horizontal travel of sand at this end.

The flattening mountain of sand is the last live dune of what once was a long system of them. Although it is so small a portion of what the sea originally produced in the area, the ridge is still massive, still an imposing surge of contours rising from the land. The dune becomes a desert in those places that deep hollows isolate. Away from the occasional valley pans of rain water and the oases of grass and flowers that grow there, curves of sand meeting the blue of stark sky press out any visible form of life—not a plant grows, not a bird passes, not a drop of water anywhere glistens to nourish the life that is not seen.

A silence of sand and sky descends into the bowls and affirms a contraction of consciousness. I forget the ridge is the haunt of deer and foxes, that the crest of the ridge can come alive after a rain with large numbers of frogs, and that the sinks are gardens then. Silence spreads itself and strengthens the aridness every grain of sand affirms. All the ponderous deserts of the world are suddenly transported to this strip of land lying so near the sea and are vividly resurrected for a moment, in this place.

Just over the rims of the deep bowls are evidence of insect life. Over a series of smooth rolls and dips of sand are vine-woven thickets where deer tracks lead. Hardy grasses able to withstand dryness and heat clump frequently between the green thickets. On the

Wilma Bradsher

west side of the dune the slopes of sand and vegetation ease gradually into the marshy shore of the Sound they adjoin. Much of the sand of the ridge is presently in the Sound, and forms long spits at low tide. Beyond the grasp of the sea, the sea side of the ridge is paralleled by a wide highway the sand does not cross. Its slopes are almost barren of vegetation and lie at a distance from the sea that is greater than that separating the western side of the ridge from the Sound.

On much smaller mountains of sand, the shore-dunes, is written the land's first dialogue with the rising sun, the sea, and salt wind in a setting of barrier islands.

Wind mimics water and etches rippling, thin bars in a meandering design. Waves of sand are caught in free fall at the curve of little hollows and bear on their crests what signs of movement have come from a tiny mammal whose feet have trailed a delicate stitching. The tracks of grackles wander and circle around the heads of Marsh Elder bushes drowned in sand and in reach of a standing bird.

In the more stable parts of a shoredune, the unobtrusive but armored Sandbur claims the ground unshaded by elder. Its thorned round fruit that is able to impale what it touches is hidden in patches of strawlike tufts. The seeds of more succulent plants plunge recklessly into the parched bowls of a dune long taproots that burrow swiftly for moisture below—it is what I sense in a sand hollow here, the struggle of land to shape itself upon tiny grains that have felt the touch of the sea.

A shoredune is born imperceptibly. A low mound builds against a natural obstruction and is fed on trapped sand, which heightens and widens to a small hill. If a period of quiescence gives the dune a temporary stability, stray seeds of hardy grasses find footholds in this margin no longer the sea's or quite yet the land's. Grasses such as Sea Oats, and American Beach Grass, with its cylindrical rather than flattened head, put forth horizontal roots, as does Running Beach Grass, a variety that while not tall by habit is able to cover sand with a creeping system of roots massed at the surface. They hold the runaway grains whose alliance with the sea is given over to a maze of colonizing roots.

Marsh Elder and Groundsel give the forming dune an interior skeleton. Propagating themselves behind the original tenuous site that began the dune, these shrubs further enlarge it; and if the dune grows high and wide enough, two aromatic shrubs become established that are not very salt-tolerant, the Wax Myrtle and the Bayberry. With a protective screen in place to shield it, a shiny-leaved evergreen, the Live Oak, can grow.

If vegetation is not a uniform bulwark along shoredunes having a short and flat seaward approach, high tides send water in that creates washouts between shrub hummocks. Water from the tidal push of the Sound can run in from behind a dune and will sometimes linger in a pan of compacted sand long enough to become a little marsh in the backdunes, which attracts egrets and herons, plovers, and other birds seeking a retreat.

61

Wind, alone, is able to carve a landscape of high mounds alternating with eroded alleys in a system of grass dunes and progressively deepens the inroads wherever hummocks are too widely spaced to secure sand. High velocity winds strip the plumes from grasses, and hurricanes shred the stems themselves into tough strands of connecting tissue, weakening the net of roots. Submergence in salt water will slow or entirely halt growth.

However, much subtler restraints are at work in the marginal zone of a shoredune. It is an hourly meeting between land and sea, repeated at all seasons in a drift of invisible droplets of salt spun from waves and falling upon vegetation vulnerable to them, that is the constant test of tenure.

According to one study, a salt droplet one-tenth of a millimeter in diameter carries thirty feet in a wind of four miles per hour if the droplet is launched at a height of five feet in the air. In stands of Sea Oats close to breaking waves, layers of salt find their way behind each lemma or modified leaf enclosing the flat inflorescences of a plume. Although a few of the lemmas below fertile florets are normally empty and have neither pollen nor pistilate structures, if salt that is showered from waves accumulates thickly enough, an entire plume is sterile, or if partially fertile, unable to complete the development of seed. The plant must then rely on its ability to propagate itself by means of lateral roots that give rise to new plants.

One particularly steep dune of Sea Oats very close to the old hotel is an anomaly in the line of low man-made shoredunes put in place north and south of the hotel to protect cottages built on the beach. The complete history of this small high dune is unknown to me. It could have been made at the time the area behind it was leveled to accommodate several cottages that were constructed on the property of the hotel. Unanchored Sea Oat roots trail down the exposed seaward face like frayed rope knotted at intervals. The "knots" or nodes are reservoirs of lateral growth. Nodes are able to root again if buried during later stages of sand replenishment, very rapidly in some grasses like Sea Oats, and to put forth and anchor an entirely new plant above the tomb of the buried one.

New plants that would indicate this steep shoredune has built itself naturally from the bottom in a complete linear progression are not evident. It is the horizontal bands coarsened with round water-sorted pebbles, shell shards, and other debris alternating with layers of sand that suggest a series of stages of development. On this eroded, nearly vertical face lie trails of dry rivulets of grains, following pronounced channels. Large pebbles have rolled in these slots for most of the eighteen-foot distance to the bottom, and without obvious cause, as I watched, small torrents of sand fell simultaneously from the rim of the dune and, fan-like, spread below it, one tiny avalanche overlapping another. Some sand was held together as a distinct ball down a channel, keeping its shape until the moment it came to rest, and with a perfectly executed sleight of hand, disappeared. Only a few of the balls were still whole at the end of a trough, having been held together by the final traces of moisture from a recent rain. When touched with only the slightest pressure, they too dissolved.

Rain can give added life to a shoredune, washing away layers of salt accumulating on the vegetation. Sometimes, unusual waves give a reprieve. On one seemingly quiet January sea creased with the light rippling that a wind makes over fields of grass, the sea's surface dotted with rafts of gulls and sometimes broached with diving gannets, or more distantly, geysered by passing whales, I saw strange waves form. A sandbar that was not very far offshore let waves roll placidly in and occasionally nudged one to breaking. If a wave crested and broke for a second time, fine droplets burst and showered the air with a screen of cold mist. Whatever angularity of the bottom that engaged the wave to form again closer to shore combined with a draft of offshore wind. In an unexpected reversal of motion, the wave's forward movement was thwarted and its crest driven back toward the sea, powdering the water behind it with a shower that fell where the wave had reemerged.

More typically a winter wave is strong enough to reach the shoredunes. The sand grains that I have watched roll down the seaward side of the dune by the hotel will be carried off by winter waves striking the dune and running parallel to its base, adding the grains fallen from the top to the ones eroded from the bottom. These grains and all those from beaches denuded of their sand by a line of dunes blocking the path of incoming waves will travel with waves that are unable to roll straight inland to deposit the sand. The waves will release their cargo where beaches are open and flat. Land will again attempt to define itself.

Is the existence of a shoredune always uncertain for the reason it is made from what is the sea's, tiny grains shaped in waves and immersed in the ancient language of restoration? Flat sand is a resilient ledger for waves: recording the undulating lines of swash one moment and the diamond-shaped patterns of backwash the next; the final swash of high tide; the veined web of sand grains fanned outward and left on newly exposed beach at ebb tide.

Offshore, sand registers the forces that raise a ridge or deepen a runnel. The abovewater stretches of sand one normally considers beach and the underwater bars that are affected by waves to a depth of thirty feet at low tide, also considered beach by those who study them, have a predictable dynamism in winter. Very large waves with short distances between crests thunder ashore then. Such a continuous surface flow saturates the berm or familiar abovewater part of the beach. Great volumes of water carrying particles of sand over the saturated berm will deposit sand where the edge of a wave stops at the top of the berm. This tends to steepen the beach face. Backwash picks up sand from the bottom of the slope, conveys it seaward, and deposits it where waves tend to break near shore, forming an underwater bar. High rapid waves in winter are responsible for terracing and narrowing the visible part of a beach and for building its invisible inshore bar.

The winter patterns of sand deposit reverse themselves in spring. Waves become more moderately sized and arrive on shore in a sequence less rapid. Sand is still picked up in backwash, but because the waves are not as large and come more slowly it is not dragged as far toward the sea as that distance it has been moved toward land. These smaller waves with longer lengths between crests starve the inshore bar. They level the terraces formed by winter waves and begin to build a wide beach.

Each year, the winter that seems so immoveable retreats and the wind at dawn is soft. Gulls stir in its currents, circling over waves. Large groups of migrant diving birds are visible as dense smoke-like clouds or as single file flights moving parallel to shore. If a line of birds is too close to shore and approaches the fishing pier near the hotel, the line is suddenly drawn skyward, as if by the grasp of a hand, then it is dropped in coils on the other side and frayed to errant strands that twist away from one another, each of them given a life of its own before merging into a single formation no longer tight as a ribbon strung out in a jet of air, but wavering loosely in the wind.

During spring migration, some of the birds that gather at the end of winter on a barrier island will head for lakes and tundra far to the north. Some will remain to nest in the marsh. Other summer residents will nest by the sea, sculpturing a simple scrape in what has become, if natural dynamisms are given play, a wide and generous berm.

If I could choose one bird to call the "sand bird" I would call the pale, ephemeral Piping Plover by this name. The bird is the color of the sand it nests upon, a ghost of the dry beach where its nest is a depression made by its body and rimmed with a few shards of shell and the fragments of plant stems. That a sand bird should be so perfectly colored for an existence on ocean beaches and on the flats behind shoredunes is ordained from the moment a downy chick emerges from the shell of the egg.

The chick is the color of mottled sand. When young are crouched away from the nest on a beach, only the thin pure notes of the parent's worried "piping," which gives the plover its name, betrays their presence. Even when the distance that separates me from an adult bird in motion is a fairly short one, the small pale bird is so instantly fading and nebulous when it comes to a stop that it is lost to my eye should my eye wander and then try to return again and shape the form. The Piping Plover seems made of sand, dissolved and amorphous until in motion.

Males have been observed making nests as part of the courting of a female in spring. As many as 20 nests are scraped in the sand. After walking a short distance the male stops, crouches, stretches forward with a pivoting motion to the left and right, and forces his breast into the sand with rapid strokes of his legs to make a depression. Occasionally the female that is following a male scrapes a nest. The male tosses aside small fragments of shells as he goes along. Sometimes he will toss them in a scrape if the female approaches closely. When a nest is chosen, four eggs are laid and are brooded by both birds.

Ocean beaches and the mudflats of Sounds and marshes serve as feeding grounds. On some barrier islands, adult Piping Plovers forage on marine worms in the intertidal zone of ocean beaches from the time they arrive in spring, through courtship, and again late in the season after young are able to fly. During the incubation of eggs and the rearing of young, the plovers move to mudflats and marsh edges to feed.

Some Piping Plovers winter here. This area of the coast is the northernmost point of the Piping Plover's wintering grounds. In fall Piping Plovers are more regularly found on dry ocean beaches than on mud flats, perhaps for the reason that their protective coloring blends so well with their surroundings and makes the birds less vulnerable to the predation of migrating hawks. By the time of a partial molt in early fall, the fledged juvenile has acquired most of the plumage worn by the adult in winter. Having deserted long before this the shallow scrape where it was hatched, it needs only to crouch and push its body into the sand with a few brief thrusts of its feet to make a solitary hollow in which to shelter. For a sandbird's autumnal sojourns by the sea, an instant nest is waiting.

5

*T*HIS MORNING A WEST WIND HAS CARRIED
hapless Ladybugs and Praying Mantises overland from
the marsh and dropped them at the edge of the sea in
long thick windrows, heaping the chitin-clad bodies
at the brink of a world neither can enter.

An odd channeling of the wind in the vicinity of
the hotel has allowed the beach here to collect a pro-
lific burst of insect life brought on a journey from
which there is no return. It is possible to know more

of the once unreachable depths of ocean frozen in darkness and fathoms of silence, and to catalog this more certainly, than what lies day-to-day by the sea's edge with its shifted tandems of life from the marsh and from the sea. Certain beaches along this barrier island, such as the one near the rustic houses first built in the area, attract the casualties of a storm-stirred sea because of the configuration of the bottom. A high tide spreads windrows as generously in those favored places as the sands are wide to receive them. Ever-changing legions are brought in that often include what has been cast back to the sea from a trawler's net or a sink net plied far offshore.

On prairies and on fields of grass, the summer harvests of hay were gathered by farmers in rows and left for the wind and sun to dry. They were called windrows. This word has come to be used for whatever is brought to land by the strongest waves of an incoming tide and laid in strands along shore. Storms make the harvests full ones, stirring up underwater troughs and sweeping shoreward forms of life that are normally unseen. They come to the sand long after a storm has passed and the sea has rested. At various times of the year floaters that have drifted on the ocean are brought in, the swimmers, the sedentary life from the bottom, the riders attached to some of this life and, from intertidal sand close by, a varied multitude of burrowers.

Some of the plants found in windrows have made a long down-coast journey. The movement of littoral or inshore currents that predominantly travel in a southern direction carry to the beach a limp, brown fragment of seaweed with small balloons or "knots" on the surface. The surface of a fragment and the knots it bears have a rubbery feel. Knotted Wrack is a visitor from the north. Shaped like a flat, branched stem that has sprouted the rudiments of leaves, it is an alga. Knotted Wrack is dislodged from the coastal waters of northern states if a stormy sea there proves stronger than the alga's holdfasts. Free-floating fronds adrift on the littoral current, which flows in a direction opposite to that of the Gulf Stream, are carried far from their original home. Olive in color or sometimes having turned a golden brown, they reach the waters of southern barrier islands.

Windrows of northern beaches generally reflect

the presence of a rock substrate. Knotted Wrack and a brown alga very like it called Rockweed grow on that base of rock, on the walls and the ledges of subterranean trenches, and on artificial bulwarks of groins and jetties, which are common in some northern states. Windrows laid after a storm or after periods of intense inshore wind cover beaches there with huge layers of the two similar algae, some subtidal and uprooted from deep water and some intertidal. The alga known as Rockweed can come on the littoral current to the windrows of low sand islands of the south, but its natural range includes warm southern waters. Both of the brown algae normally have numerous air sacs on their fronds. These conspicuous spheres are solitary in Knotted Wrack, which is without a midridge on a frond. In Rockweed, the sacs are arranged in pairs along a distinct midridge.

So different from either of those brown algae are the few ribbony green strips that are among this morning's strange windrows of insects. The glistening strands were not torn from low hills of sand lying under a blanket of waves, as Wrack and Rockweed are torn from rock in the north. The beach by the hotel will gather the vegetation that a steady southwest wind pushes from the marshes through the inlet that lies south of it, and the ribbons as well as the insects are from the west. Of the widespread group of plants with flowers and an intricate interior structure, the plant to which the long leaves belong is one of but two that can grow submerged in salt water. The linear leaf gives this aquatic the name Zostera, meaning "belt." The inner cells in a leaf are in tiers thick enough to prevent the outline of an object placed behind a leaf from showing through, and as many as five veins of conducting cells run the length of the yard-long ribbons.

Mats of whole leaves at the final stages of their growth, blackened and having died back naturally, fragments broken from living leaves, and those deliberately cut by an animal of the marsh resembling a muskrat are laid by littoral currents driven by a southwest wind on beaches north of the inlet. When I came to the hotel and the edge of the waves on the first of my visits, I was unaware of the journey the leaves had made, and this flowerbearer drew me to the new and unexpected treasures of the sea that its ribbons enwrapped.

In its natural habitat in the marsh and Sound, Zostera has numerous leaves sprouted from a thick carpet of roots able to hold and bind the fine sediments settled out on the bottom of water that is relatively undisturbed. Broken leaves are often found in the windrows no wider than what a six-inch tide spreads on small, crescent-shaped mud beaches in the marsh.

Widgeon Grass is the other flowering plant brought through the inlet from the marsh to the beach. Like Zostera, it is able to flourish while submerged in its natural habitat. (*Spartina alterniflora,* the most salt-tolerant of the cordgrasses, is also a flower-bearer, but it must grow above water.) Widgeon Grass is named for a duck, the Wigeon, that feeds upon it in the marsh. The leaves are thin and hair-like and do not show veins.

Widgeon Grass and Zostera construct the underwater nurseries in estuarine waters and form the protective cradle for sea-started fish, which as "fry" migrate from the sea to quieter waters. Before returning to the sea the fry grow and by midsummer move down to more open stretches and the wider forests of these two aquatic plants.

Like Zostera, Widgeon Grass is an anchorage for small animals with shells and without (and for marsh algae needing another plant upon which to grow). The warm waters of the south are home to the filamentous Red Algae that resemble Widgeon Grass, such as Hooked Weed, Agardh's Red Weed, and Graceful Red Weed. Hooked Weed is not always hooked at the end of its branches, which are shorter and denser overall than those of Widgeon Grass. Agardh's Red Weed has its branches tappered at their bases. Graceful Red Weed has a compressed look. Certain Red Seaweeds such as Banded Weed, the small Barrel Weed, and Tubed Weed have their branches respectively banded, beaded, and divided crosswise and lengthwise with fine partitions. The same pair of binoculars that gives detail to far off objects will, if reversed in the hand, act like a microscope and verify these distinctions.

Arising from the kind of horizontal, thickly intertwined network of rhizomes that Zostera has, Widgeon Grass stabilizes water habitat that tends to attract sediment. Whereas the flowers of Zostera are stemless, small, and invisible behind the bases of specialized fruiting leaves, and open in succession on a central axis that stays concealed, the flowers of Widgeon Grass grow a slender stalk at the base of an ordinary leaf and open simultaneously on a flat column. The stalk eventually carries the fruit of Widgeon Grass to the surface, and a floating cluster of nutlets becomes visible.

Like Zostera, Widgeon Grass is found on crescent mud beaches in the marsh. The abbreviated drift lines of what a six-inch tide wreathes in the marsh along openings in the cordgrass, or those a "wind tide" forms—the push of a prevailing wind—contain little life other than plant fragments. Derelict docks, the hulls of sunken boats, pilings or posts, submerged structures of any origin attract sessile creatures seldom pried from their sanctuary.

A "wind tide" that builds on a strong steady blow across open shallows might edge to a mud beach a shell as long as my finger and almost twice as wide. Square ends, an oblong shape three times longer than broad, and a hinge formed by two tiny teeth half way down the length distinguish the Stout Tagelas.

Marl discolors the shells of various kinds found

in the marsh. They will remain unsculpted there, where the sound of waves is not heard on a tide of wind.

On ocean beaches, numerous and varied are the finds of most any morning's search for life of the sea if I open windrows before the microcosms preserved in mats of vegetation are lost to hungry gulls or to grackles that have learned the mats are repositories for food. Ghost Crabs hunt them, the pale secretive shadows sculling over the sand at night and, by day, the tiny beach scavenger keen as a Ghost Crab, the Tiger Beetle.

Some mornings, as if to invite an unlocking of troves strewn by a high tide, the edges of a matted forest catch the downy shaft lost by a gull or tern, or the slim and sharp feather of soft buff, strong black, and pure white that is a shorebird's; or the well-oiled feather of an Osprey. An early disentangling of fresh windrows can reveal little starfish, the white, papery sphere of a cluster of egg cases shaped like a golf ball, even the rubbery branches of a Bryozoan that might be mistaken for a coral or an alga or a sponge, but that is actually the house of many individual animals called zooids.

The fast-fading fragments of animals found only in the sea, the algal life known commonly as seaweed, the stiffly branched and brightly colored Octocorals, the grapelike objects that are animal life and not plant life, the shells difficult to identify, those easily named, all are indiscriminately horded. Well-domed, mostly oval bivalve shells that lack at their hinge the two symmetrical "wings" or flares a scallop has and that possess numerous ribs raised in a widening fan from a point on their hinge are either Cockles or Arks. Fine linear teeth near the curl of a fresh shell are visible upon the interior rim if an Ark Shell; absent, if a Cockle. A bud-shaped white shell about an inch in length on a long, thick, dark stalk that attaches it to a piece of driftwood or some other substrate, whose folds of shell are edged with rich orange and pale blue, is a Pelagic Goose Barnacle. A passive floater, it is apt to have traveled long distances at sea.

From a few yards out from the shore in ocean-churned sand comes the robust but thumbnail-sized Coquina Clam, painted yellow, red, or purple if fresh, and banded, or rayed.

A shell about an inch long with a yellow or orange gloss, almost transparent, gives out a pleasant metallic sound when brushed by another shell. The musical oval is a Jingle Shell. In its growth the smaller, thinner bottom half of the bivalve assumes the imprint of the rough and irregular surface to which it binds itself and has the look of having been damaged. A noticeable break at the anterior end leads whoever finds this half to mistakenly believe it has been accidentally notched. But the opening is natural and always present, allowing strong byssal threads to emerge and bind the living shell to a mooring of rock or of wood.

A shell shaped like a cat's paw and named for the resemblance arrives either free or cemented to another shell.

A light gray shell several inches long resembling the head of a parrot without a beak is a Moon Snail.

The whorl on the upper surface of the shell ends as an "eye." If the snail is still inside, the shell's large round flare will be tightly closed by the hard, horn-smooth foot of the snail. This durable, impenetrable door to the shell can be found by itself in windrows and recognized because it resembles the shell's outer surface. The hollow core on the underside of the shell is the umbilicus. In the Northern Moon Shell, this indentation on the lower surface is quite obvious. In the Lobed Moon Shell, the more common one found on southern beaches, a dark callus fills in the umbilicus.

Lacy and fernlike, fanlike, bundled in tubes, or banded, and all usually small are the algae that assume a conditional life and come to ocean windrows attached to larger algae, just as some land plants will grow upon the surface of others. Some of these are crustose and brightly colored, some are mucilaginous; some have the texture and habit of land plants. Very delicate algae deteriorate after a few hours of exposure, and unless the search for them is undertaken in the early hours of morning before the sun has had a chance to desiccate the fragile tissue, they leave no trace of themselves.

Always to be found in ocean windrows are the fragments of the shell of the Mole Crab, or the walking legs, claws, or body segments of the true crabs. Mixes of individual chitin parts that do not belong to any one animal make a chimerical creation when discovered in the shape of a perfect ball. The small, light-weight conglomerations are likely to remain a mystery to a visitor, for the chance of witnessing the manner in which they are left on a beach is not very great and, if seen, the process can be easily misinterpreted. The balls are the undigested parts of animals eaten by shorebirds and expelled with the kind of motion a bird would make if food were too large to be swallowed and caught in its throat. Terns and gulls and skimmers will leave the perplexing collections. Homogeneous balls of tiny whole shells such as those of the Coquina clam, balls of fish scales and fins, sometime with bits of plastic and metal, linger in windrows.

Green algae cells called phytoplankton fill the waves of the sea with "bloom" in spring and in fall, even when waves seem barren, forming at those times luxuriant windrows in their own minute realm that go unseen. Similarly elusive are the windrows left by the passing of large schools of fish, the "slick" that is loosened by the friction of countless bodies pressed close and brushing one another as a school migrates.

Discs of bone, sometimes small and sometimes large, can be found in windrows. They are the strong and durable plates that cover the gills of a fish.

Objects of human origin litter the sand on any one day. Tangled pieces of net might come from a pier, from a shorenet destroyed by storms, or from one of the many fishing boats that pass offshore, or from a net laid in the Sound. Pieces of wood and plastic are nearly always recognizable, even glass roughened and clouded by tumbling in the waves.

Not all the objects made by human hands are easily identified, for the crinkled, toughened houses of sea worms and the egg collars of moon snails, both found more often in fragments than entire, can be ambiguous. The tube of a sea worm resembles the crumpled portion of a paper straw. The flat circular base holding the minute eggs of a moon snail seems to be a piece of cardboard until it is picked up, and it disintegrates to the fine sand of which it is made.

More surprising is the total metamorphosis of the Portuguese Man-of-War, Physalia. Whenever the circulating winds of hurricanes travel up the coast, they bring this tropical jellyfish along with other travelers of the Gulf Stream to ocean windrows. Living, its beautiful sail is raised on its float and rimmed with neon pink. Windrows still freshened by waves are the vase of this venomous orchid of the sea. Beneath the sail, Physalia glows with a solid electric blue at the attenuated end of its air-filled float and, in the interior, with the blue of glacial ice, a blue that deepens gradually along the perimeter of the float. Vivid bottle-green lights the blue. Below the float that slowly contracts and curls when the animal is stranded hang the poisonous tentacles of brilliant blue and purple.

During the hours the sea touches it all colors hold true. When the sea withdraws they fade like a flower's. The broad sail rising vertically from the top of the float pales and sinks to a mere ridge; but the float itself becomes hardened and rigid with air. Thus transformed, Physalia can linger for days on a beach. Pressure from my unguarded step makes an audible, sharp crack at the moment the float explodes and its membrane flattens—the transformation completed that has changed Physalia to a scrap of cellophane.

The clean, bony jaws of a fish, the empty carapace of a Horseshoe Crab, the pure white breastbone of a gull after weeks of drying are able to hold the distillation of sea and to release the scent if moistened again.

The strong scent of sea is in the bushy fragments of Sargassum that come by way of a southeast wind. Like Knotted Wrack, Sargassum or Gulfweed is golden brown when fresh, and it bears round, tiny, inflated sacs like Wrack's or Rockweed's. The resemblance is only superficial. Gulfweed's sacs are on short stalks, and those of Wrack and Rockweed arise directly from the surface of the plant. Gulfweed's "leaves" have toothed edges, Wrack's and Rockweed's are smooth on their margins.

Of three species of Gulfweed, the two that are free-floating drifters are readily distinguished from the rooted one, which bears small dots on its leaves. Of the two drifters, the more likely to be carried to land is *Sargassum fluitans*. The alga is not uncommon in windrows of this particular part of the east coast.

The other species of floating Gulfweed has a long projection at the top of a sac. A sprig of Gulfweed appears to hold no further interest beyond its origins in a distant floating sea of plants. What a wealth of life would pass from my eye if I returned a fragment to the sand without a closer examination.

A single bubble-like sac, or any part of the surface, becomes a lesson in the profusion that waits in the sea to begin life; and the reversed barrel of a pair of binoculars reveals the growth of this life. Any fresh fragment might hold the white, coiled tube of lime that belongs to Spirobis, the Hard Tube Worm. The coil is no larger than a salt grain. Close by might lie the precisely geometrical filigree that is the countless boxlike chambers of individual animals, the zooids of a Lacy Crust Bryozoan. The scale is so incredibly small that the white cast seems but a fine powder until the aided eye sees its delicate structure. One species of Lacy Crust Bryozoans has a knob on both upper corners of each box. That particular Bryozoan is often found on Gulfweed, and its species name refers to the two knobs.

A Gulfweed sprig will grow rampant vines of

brilliant red and fanciful gardens of cup-shaped "flowers" on tall, ringed stems. Each flower is the house of an individual animal or zooid, quite different from the zooid of the Bryozoan that occupies flat, lacy chambers. The animals are Hydroids. Hydrozoan species at the zooid stage of their life occupy structures resembling flowers to such a degree that their names are derived from the similarity. Even the naming of parts of an individual zooid is botanically linked.

The lavish gatherings on certain beaches rather than on others, the sudden appearance of a trove of findings after a succession of tides that have left little or nothing on the sand, and seasonal bursts of life from the marsh or ocean whose exact rhythms are unpredictable entice me toward one more landmark in the distance; around one more curve of sand, around one more terrace.

The fall sea offers just a few spare and undulant tracings of a night's high tide, should I fail to remember to go closer to the dunes than in spring or summer to search windrows. The autumn sun strikes sand in a certain way, and one can know by looking at how the dunes absorb the light what season it is, as if a heaviness in the sunshine lets light sink into and saturate them. Aster and Ox-eye Daisies bloom by the roadside and leeward of the dunes, and the wind sways the flowers as it passes toward the sea over sand that no longer thins and reflects light. On the seaward side of dunes, Cakile the Sea Rocket grows. Its leaves are thick and spongy. Oblong, often several inches long, they narrow at the base; their margins bear broad and shallow teeth. Small white or purplish flowers have four petals arranged as a cross of four arms that are equal in length. The plant grows close to the water, for its leaves tolerate submersion in stormy seas. I have even found its flowers still blooming after a night of heightened waves.

When this flowerbearer meets the edge of autumn waves, it is able to enter the world of the sea. A floating watertight fruit easily given to the sea in the

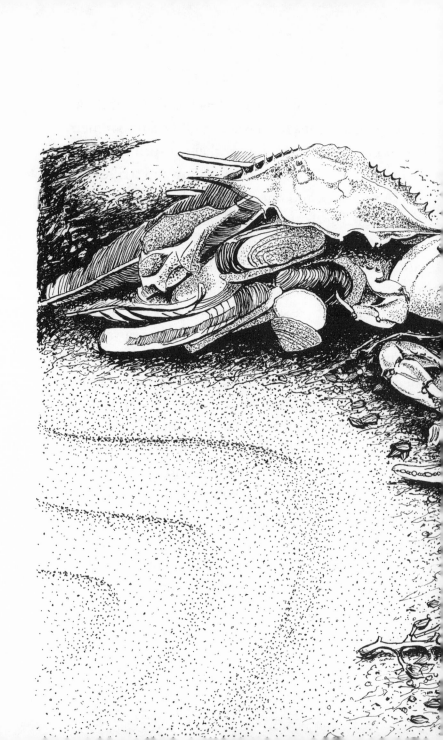

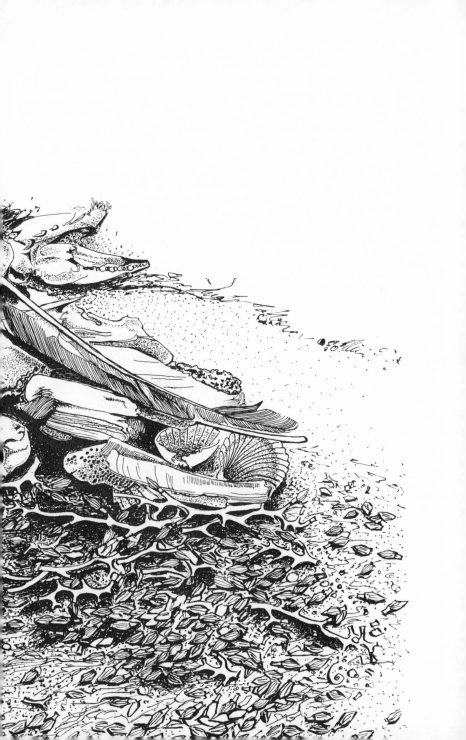

fall makes Cakile a dweller of two very different worlds. Unique among flowering land plants with a high tolerance of salt spray on ocean beaches, Cakile possesses a fruit capsule with a central constriction. The adhesion of the upper segment to the lower one and that of the bottom half to its stem is very weak when the fruit is mature. Neither half has a seam along its length that allows it to split open and release a mature seed, as the capsules of other seed plants do.

This design has given Cakile a way to float the fruit along the autumn waves. The upper portion of the capsule always contains a seed; the lower is sometimes empty. Because of the corky layer in the upper segment, this part travels a longer distance, and after drifting far from the place where the flower that produced the seed has paled and died, it comes to lie in the windrow of the tide that courses highest up the beach. The seed germinates and takes root in the sand. From what the sea has returned, a new generation will gather in its leaves the sunshine that matures its fruit and, in some species of Cakile, turns it an autumn gold.

Of all the bits and pieces of life brought to the

94

windrow of any night's tide in fall, a large claw fresh and bright red is the mark of the she-crab Callinectes. Its presence attests to her lone, purposeful exodus that she has made from the west to lay her eggs in the sea. A female bears the red color for the duration of her four to five-year life span as if signeted for her journey. Normally the home of Callinectes, the common Blue Crab, is in the marshes, Sounds, and bays where the right amount of salinity exists due to a mixture of fresh water from rivers and salt water from tidal flow through an inlet. There the she-crab lives most of her life, there she mates after the most difficult molt of her shell.

She-crabs have a dome shape on the central part of their underplate; in males the dome is compressed. Sometimes, young males have a bright red wash on the ends of their two fore-claws, but males have on the underplate of their shell a pattern that is different enough to always distinguish them. When winter comes, males and those females that have not mated seek the deep channels of the Sound in which to shelter. It is not a trick of the ebb tide through the inlet, or the force of a sudden overwash, or any chance wind that carries the older female to underwater ocean troughs in the surf zone. Bearing the wash of the sun on her claws, she follows an obscure purpose at a time that is perhaps richest in some ancient wisdom and makes her passage to the sea.

Part Three

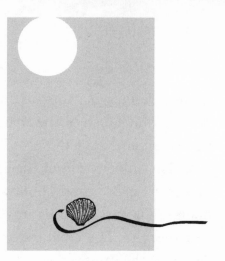

6

*O*N MY WAY ALONG THE EDGE OF THE
sea to look for shells, I come upon the entire carapace
shed by a crab found only in the sea, and I am brought
to think of the strength of fragile things.

The details on the margins are in place; grace-
fully tapered "wings" that widen the top beyond the
stretch of my opened palm have not been cracked or
chipped by waves periodically skating in and carrying the carapace along the beach. Winding labyrinths

in the interior where organs once were housed are still perfect, and much of the structure of the abdominal plates remains intact.

The smooth top of the carapace bears a pattern of distinct pebble-sized hollow ovals of faded, tannish orange. The largest ovals are on the wings and by the rear margin. The smallest are crowded toward the center of the carapace top, giving the background of creamy yellow an overall speckled look and the crab its name, Speckled Crab.

The protective shells of crabs are made mostly of "chitin," a Greek word meaning outer garment or tunic. Before the evolution of bone, this was the only substance strong enough, other than that of hard phosphatic shells or that of calcareous shells, to anchor muscles, to shape the body, and to protect soft tissue. Extinct forms of joint-legged animals called trilobites have left traces of their chitin carapaces in the sediments of ancient Paleozoic seas. Chitin also forms the outside skeletons of insects. Crabs, insects and trilobites belong to a group of animals called arthropods. Only arthropods possess jointed legs and a segmented body.

Legless animals that produce shells of lime typically thought of as seashells are mollusks. Mollusks have unsegmented bodies and a specialized "pseudo foot," distinctive of this group. Their shells of calcium carbonate are not molted like the shells of arthropods, but are gradually added to as the animals grow.

Unlike the hard lime shells of mollusks, chitin can be jointed to form flexible walking legs and functioning claws, gills that breathe, and a mouth. Made mostly of organic molecules, it is found not only in the shells of crabs, insects, and trilobites. It spans two kingdoms of living things by composing the cell walls of some fungi.

Shells of the common crabs found on the beach are a mixture of chitin and the calcium carbonate utilized by mollusks. Calcium carbonate increases the strength of a crab shell. Periodic growth involves a process totally different from that which produces the extensions of a mollusk shell. A crab must actually escape the old covering by secreting a new one that is larger and able to accommodate the animal's increased size.

The crab does this by producing an enzyme that

dissolves the innermost layer of its old exoskeleton. A shell that is at first quite soft forms beneath it. The outermost shell splits between the abdominal portion of the carapace and the top part, the shield. The animal shrinks slightly, backs out and leaves a cast of itself that is complete in every detail, including the facets of its eyes and the lined parts of some organs. Slowly the animal's new and larger shell hardens and becomes rigid.

Sometimes I come upon a crab that has just been brought in and is still living. But it is not often. Or do I often discover an entire carapace that has been shed. It is the single top plate or shield that I commonly find on the sand. Because of the natural and periodic duplication necessary for growth, inhabitants of the sea that stay hidden much of the time give up something of themselves that otherwise would be much rarer; I have the chance to wonder over the delicate touches on the "beauty box of Persephone" and all the orange, oval runes it bears. These tiny circles decorating the unusually shaped crab named Persephona resemble those on the shield of a Speckled Crab.

Persephona grows to two inches and is different in its proportions from the winged, shallow-domed Speckled Crab. The genus name Persephona, which is the Latin rendition of the Greek "Persephone," comes closer than the common name of Purse Crab to suggesting the elegance of the shield's high dome, detailed along the margin by a row of delicate "beads." The story of Persephone's beauty box is from Greek mythology. The box was to be carried to Venus to redress a grievance and was to bear within it some of the radiance of Persephone, daughter of Ceres and Jupiter. Irregular groups of small, loose ovals flank a pale yellow, narrow line at the top of the shield. The ovals and patches are purplish brown on a shield freshly shed and fade to tannish orange on older shields. Individual ovals break away from the clustered pattern at the top of a shield, decorating the sides near the beading along the rim and near a "hinge" of three short spines.

The ovals in the design of Persephona and the Speckled Crab are examples of a surprisingly tenacious theme of color and pattern in the shields of crabs I find on the beach. Ovals decorate the shallow shield of another, more common crab, the Rock Crab. At a glance, its design is nothing like the pattern of either Persephona or the winged Speckled Crab. The few plainly visible oval shapes on the shield of the Rock Crab are actually the places where an otherwise heavy and even accumulation of minute reddish brown dots are absent—but the dots themselves are hollow ovals, much like those of the Speckled Crab, and so reduced in size that each seems solid to the eye.

This fine stippling packed upon most of the surface of the shield of a Rock Crab is shared by a crab very similar to it, the Jonah Crab. The Rock Crab and the Jonah Crab grow to widths over five inches. Often the margins of crab shields are toothed, and the number of these marginal teeth, which includes the tooth of the outer rim of the eye orbit, can be used to distinguish crabs that look alike. However, a count of marginal teeth, nine, will not help to separate the Rock and the Jonah Crab. It is the shape of their marginal teeth that is diagnostic. Those of the Rock Crab are smoother and more even. Those of the Jonah Crab have smaller secondary teeth that makes a tooth appear jagged and rough. The wingless shields of the two crabs are broader than long, and the ovals in their design so numerous that they give a reddish brown color to the surface.

The intricate similarity of the shields escapes the unaided eye. Only with a hand lens or the reversed barrel of a pair of binoculars does the oval design of both the Rock Crab and the Jonah Crab show an affinity to that of the Speckled Crab and Persephona. The ovals that spell some common language of the sea are written more persistently on the most fragile of all the different shells sharing it. A full play on the theme is revealed on the thin shield of a Lady Crab. Decorating its nearly circular and wingless shield are minute ovals no bigger than the dots of the Rock and the Jonah Crab. The distinct center of each oval and its colored rim are discernible with magnification, as they are in the Rock and Jonah Crab. But in the Lady Crab the tiny ovals are clustered to form larger ovals, whose shape is irregular and sometimes interrupted by the shield's cream-yellow background.

The Lady Crab grows to a width of three inches. After a shield lies on the beach for a while, its pale gray background fades to creamy yellow; the purplish brown of the rim of the ovals turn a reddish to orangy tan. A Lady Crab's pattern of large ovals, and a count of five marginal teeth identify her. In summer, shields smaller than three inches are common and are very breakable. Those that I am tempted to gather and slip into a pocket so that I might in a quiet moment on the hotel's veranda try to decode their rune are seldom drawn out whole again.

Carapaces that are regularly shed as the natural method of a crab's growth make it almost certain I will find shields. Even a beach with a particularly heavy summer traffic of strollers will usually have them lying somewhere on the sand. The slightly smaller duplicates of the armor protecting a crab's soft body on the upper surface are weaker and more brittle than what replaces them in the living crab. Perhaps the fact that a carapace is no longer the heavy material it was once is the result of the dissolving process by which a shield is cast off and abandoned to the sea, combined with a further leaching of the calcium carbonate composing it. Or perhaps the fragility of a shed shield comes from its drying out on a beach and the ensuing and progressive flaking of thin layers from the undersurface.

The empty cover of organic chitin, heavier, more durable in life, that functioned to protect the soft body of a joint-legged animal is not ordinarily thought of as a sea shell. Yet it is of the sea, it is in part made of lime, and its shape and decorative design can be as intricate and as puzzling as that of any shell of lime.

The shells composed wholly of lime and grown by the animal known as a mollusk most often bring strollers to the sea's edge to find the perfect and fresh shell. Small sea shells buried in piles of thick wind-rows that are the aftermath of a storm will still be miraculously intact and will reward the thorough searcher who takes the time to investigate those wind-rows. Houses of lime in all proportions and sizes are the shelters of the legless animals that either live a somewhat sedentary existence on the bottom of the sea or that hunt it regularly, preying upon other mollusks, or that attach themselves upon the surface of a floating object.

One particular mollusk, perhaps the most inventive of them all, sails its shell home alone, upside down below the sky and above the bottom of the sea. The Violet Snail lives on the wave tops. Observers at sea have looked for its translucent house in vain. Suspending itself as it does by a "float" of bubbles that keeps it from sinking, it is too much like the froth of the sea to be detected as it wanders along the edge of the Gulf Stream. But an east wind sometimes diverts it and ends the odyssey that it had made while hung from its airy pontoon. The inch-and-a-half shell with its ramp-like flare where the bubbles are held is sailed to shore, the violet globe perfect. I have searched for, without ever finding, this small sheer shell that others have come upon not very far from the old hotel.

Winter seas lavishly give up their treasures of much sturdier and more common shells. Months later near the base of the dunes south of the hotel, the remnants of hordes of shells brought to shore by winter storms can still be found. Debris that high seas has swept along is also lodged in these places. The evidence of how far waves are able to travel surprises the visitor who knows only a peaceful summer sea. Shells, wood fragments of the hotel's walkway to the beach, splinters of broken cedar shakes, broken branches of Marsh Elder and the strawlike stems of Sea Oats are all gathered under the barricading line of dunes.

When a summer beach is without large conical shells known as whelks, these old windrows laid close to the dunes can hold them. The perfect shells brought to shore by winter waves have already been gathered from the open beach. In summer, I usually find only those worn by sand and water; often they are broken by the tires of vehicles that have passed along the beach and further waterworn after becoming fractured. Normally a whelk has a wide flare; the body of the shell rolls into a tight spire. Conspicuous protuberances or knobs that are placed evenly on the shoulder of a shell, and continue in diminished size near the spire, give the Knobbed Whelk its name. If held upside down with the spire at the top, the flare of a Knobbed Whelk widens to the right as a typical whorl in a whelk shell does. A deep square-bottomed channel winding into the spire from the shoulder is

distinctive of the Channel Whelk. Should a whelk shell have a flare that widens to the left instead of to the right, it would be a Lightning Whelk.

The living animals that inhabited the numerous Knobbed Whelk shells I find in old windrows were twisted through a 90 degree angle during the formation of their whorled houses. Instead of the radial symmetry that a creature like a starfish possesses, by which body parts are arranged like spokes in a wheel, mollusks tend toward bilateral symmetry, by which parts are paired on either side of a line. Some biologists think that the organs of a whelk are originally paired in this way, and a progressive displacement reduces the size of one in a pair within the animal's shell. The gradual twisting shifts the cavity between the body and its growing part, the mantle. The mantle is forced upward and concentrates the nervous system in a head. Growing only from its anterior end where the mantle fold contacts the lip of the shell, the animal takes on the unusual form of its shell's interior.

The Knobbed Whelks stranded beneath the dunes are variously sculpted by having been tumbled and scoured in sand-filled waves. Should the close inner folding near the spire of a shell become detached and the spire planed away, this oldest part is transformed to the rare "rose" of the sea found every now and again in the sand. In some of the shells, the flares have a wide window through which a twisted inner column is visible. In others, the flares are almost gone. Waves have explored endless themes in paring the parts of a shell weakened by fracture to only a suggestion of what was once there. The interior column is left as an abstract composition of the simplest lines of pale orange, flesh color, or gray that is mottled like marble, as if an unknown artisan has tirelessly worked to solve the riddle of mysterious form.

Along with the Knobbed Whelks that have come to lie on mats by the dunes are strands of their egg cases. All the tannish sacs fastened to the strands are blunt at the edge; egg cases of Channeled Whelks resemble them in color and texture, but are noticeably tapered. The strong, parchment-like sacs bear a small central opening at their end through which large grains of sand have been forced, and they rattle if shaken. Each opening is a door by which several dozen young whelks made their escape. Among the grains I spread in my hand from one sac, I find a perfect miniature shell. In another sac, all the "grains" are shells. Sealed there by some misfortune, taken from the grasp of the sea, such shells are the tiny replicas of what they would have become if carried through the inlet to the quiet waters to the west, and would have remained, far from the sculpturing fall of waves.

Not all species of mollusks produce shells. Those that do draw calcium ions from the environment. The electrically charged particles are transported in the bloodstream and are then secreted in the cavity between the body wall and the mantle. The mantle is the covering of tissue giving rise, in the larva, to the hard lime shell and to the additions made to it in the course of the animal's growth. Crystals of calcium carbonate precipitated from the outermost cells of the mantle are positioned parallel to one another, but stand perpendicular. This upright placement in the shell wall gives the shell its bulk and strength. In the case of the thin pearly layer that lines the shell's interior, the crystals of calcium carbonate are also parallel to one another, but lie flat, a prismatic layering that gives the shell its smooth luster.

Very early in the history of life, animals were producing shells of calcium carbonate. Fossil mollusk shells six-hundred million years old have been found in rock that is the hardened sediment of shallow Paleozoic seas. The fossil record of mollusks is without interruption from their appearance in these ancient seas to the present.

Geologists have assigned to the history of life three great divisions of time. Each division is determined by a change in the representation of major fossil species. The three divisions or "Eras" reflect significant advances in the complexity of vertebrate fossils found in rock. With comparatively little change, mollusks lived through the Paleozoic Era that saw the beginning of bearers of bone and through the Mesozoic Era when reptiles ruled and into the Cenozoic Era, the age of mammals. In a geologic time scale reckoned by millions of years, the Cenozoic Era has only just begun; the rock sequences of the Cenozoic define the shortest and the most recent sections of geologic time, its "epochs." Non-marine rocks of the "upper" Cenozoic hold the unfolding record of the human story. The rocks that are marine in origin contain species of shell-bearing mollusks found alive in modern seas.

In the 1800s, names for the epochs of the upper Cenozoic were proposed based on the percentages of these still living mollusks found in the marine rocks associated with a particular epoch. Numbers of currently living species of mollusks diminish in proportion to the position of a rock layer; the lower the layer, the fewer living species are found.

The Miocene Epoch is taken from the Greek "meios" meaning less; only 20 to 40 percent of mollusk species found in modern seas existed then. The next epoch, the Pliocene, from the Greek 'pleios' meaning more, has 50 to 90 percent of living species. The current epoch, the Pleistocene, from the Greek "pleistos" meaning most, has 90 to 100 percent. The classification has not proved to be a precise one because of the tendency of the distribution of shells to be so unpredictable from place to place. Yet the names have settled into geologic literature and linked our human origins with shells.

I have puzzled over the symmetry of opposed halves possessed by the most common clam of the beach, the Marsh Quahog. One valve of a pair does not duplicate the other and is not transposable with the other. The same bilateral symmetry has hallmarked the bearers of bone from their advent as little swimmers in Paleozoic seas and through to every stage of complexity.

The animal inhabiting a shell of two valves that are opened and closed by specialized muscles leaves scars of these muscles on both valves. Flanking the umbus, which is the hinge region of a valve, are the shallow banded impressions resembling a light thumbprint where muscles were attached. Connecting these scars is a distinct curved line. It follows a Marsh Quahog shell's perimeter where the shell begins to flatten. This line is the place where the folds of the actively growing mantle were attached when it was producing the extensions of both valves. A wash

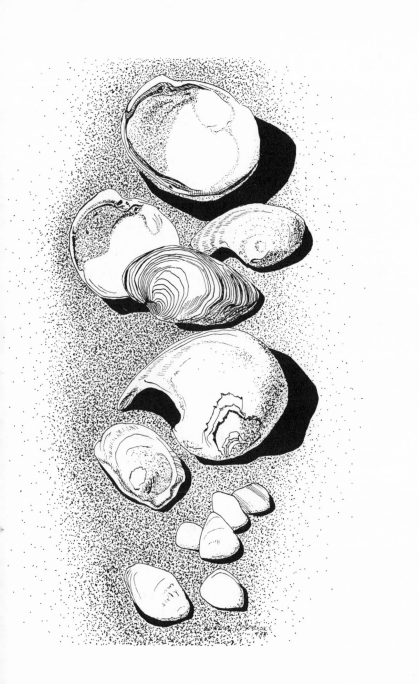

of purple sometimes follows this line around or seeps deeper into the body of the shell in the region of a muscle scar.

When the animal is alive, its two shells have a thin brownish or blackish "skin" camouflaging them. This outer organic layer, called a periostracum, is grown by the animal over the surface of its four-inch valves and blends with the surroundings. The skin disintegrates if a shell lies on shore. If the shell is laid high enough on a beach slope, it becomes wrapped in a bleached "skin" of solid white that covers both the interior wine-stained surface and the exterior. A bleached Marsh Quahog on a beach barely attracts the eye, and nothing further holds the eye beyond an initial, cursory glance. Shells of the Marsh Quahog are so prevalent along this part of the coast that north and south of the hotel whole stretches of sand are filled with shells whitened by the bleaching effect of the sun and made ordinary.

Whenever waves come high up the beach for a long enough time to succeed in scouring the shells, this camouflage of white is worn away. If the shells are again exposed to the sun, the wrapping is renewed.

Not far from the remnants of the Knobbed Whelks brought to the dune base by winter seas are long swaths of these bleached Marsh Quahogs. Some of the shells in the swaths show their plum-colored stains where the edges have been worn. Shells chipped by the tires of vehicles riding over them are the easiest to identify. On the beach are shells similar in shape to the Marsh Quahog. The Ocean Quahog grows larger, but lacks a purple wash along the interior; a two-inch shell called the False Quahog is colorless.

During my first summer by the sea, I confused different kinds of quahogs with bleached Marsh Quahogs and did not realize so many wine-colored shells were wrapped in white. The railing of the hotel's walkway to the sand or that of a veranda, or one of the low round tables of juniper wood usually had on it a shell of the Marsh Quahog. The Algonquin Indians that once lived in this region used the Marsh Quahog as a medium of exchange. The shell is sometimes called the "wampum shell." An early account of the Marsh Quahog and its value as currency records that the deeper the wine of the interior, the greater the value. The name Mercenaria as well as the common one of wampum shell refers to the use of this quahog as money. Despite its other common name of Marsh Quahog, this mollusk can be found living in the sea as well as in the marshes; it was even more plentiful in the 1700s.

Its use as wampum did not involve the whole shell, as one might suspect. A shell half was fractured by the Indians and the purple fragments drilled through by a laborious and skillful process. These shards with holes were ground upon a rough surface to the shape of small beads that could be strung together and given worth according to the length of a string. Only the simplest of tools were used. A nail was placed by an Indian in the hollow of a strong reed, and the nail rotated by rolling the reed along the outside of the upper leg with one hand. The other hand anchored the shell fragment to maintain a firm and steady contact with the nail end.

A shell of the Marsh Quahog is extremely hard and difficult to work. Efforts to drill and shape a fragment into a bead using only the most rudimentary tools had been perfected by practice and patience to produce what was called "peak" in North Carolina.

Traders attempted to duplicate the working of peak, but they found the Marsh Quahog too difficult to drill without fracturing it and too tedious to shape and smooth to a gemlike luster.

The sea, too, is practiced and patient. Like sculpted Knobbed Whelk shells shaped and smoothed by winter tempos, the shells of the Marsh Quahog take on a new existence when long swaths are in reach of the fall of waves. Among the valves are many small shards that are flat and round. They are the fragments of broken shells. At the edge of waves their bleaching has been washed away; they are wine-stained and scattered liberally in a swath. Some of the ovals are banded and are the fragments from the region of a muscle scar. Other fragments are solid burgundy. Some swaths have shards whose purple is richer and darker. They are the fragments of shells that have undergone temporary burials in mineral-laden sand. A careful search discloses the presence of strange round "seastones" that are marbled with gray. The seastones are a shell's hard and durable umbus pared of its flare. Their marbled surface is burnished by sand-filled water to a silken luster and by a layer of spindrift is given the gleam of a semiprecious stone.

On yet another part of the beach near the hotel is a swath of Marsh Quahogs whose burgundy shards are the richest of all, the rare seastones marbled most deeply. Dune-cuts nearby show dark horizontal layerings of peat rather than of sand. The outcrops of peat reveal that the beach was originally part of an estuarine labyrinth, which the sea has reclaimed, and these shells are from marshland. Now at the sea's edge, they are minted like a currency, mutable in a world of waves whose sounds are the mock codas of unfinished things.

7

*I*N THE FORMATION OF THE EARTH FROM the material of the sun, heat from the pressure of gravity allowed the flow of dense molten rock toward the earth's core. Lighter material surfaced and became the granitic rock composing the outer crust. With the intense compaction and heat generated by gravity as the earth took shape, enough water was released from the dense compounds at the core to fill the first seas.

Material that was originally the sun's continues

to be a source of water in modern seas. The coast of the Atlantic reminds me of this primordial tie in that it is here the sun is renewed each day over waves. If I should leave the edge of the water to climb to a nearby terrace for a better view while color is dropped suddenly into a gray sea and floated just beneath the surface within a wave trough, the fire sliding backward into a farther trough and then into the next until a torch of lava extends from the horizon to rosy sand, the effect can be lost with my change of angle. Each sunrise is not like the previous one, and it will not be like the next. The position of the observer, the shape of the beach, the moving clouds, the succession of equinoxes and solstices, and the shifting waves influence each coming of the sun.

Today at dawn a deep rumble hangs against the backdrop of sky, behind the multiplication of waves falling upon sand. The wind gathers this immensity of sea in a billow of sound, tunnels it to the ear one moment, and changes it as suddenly the next to the dead draw of a vacuum. The first of the gulls come out of the thin light, their dark forms passing over the curls of waves and wide sweeps of foam. As the light of the hidden sun grows stronger, gulls fly more frequently along the sea's edge. Soon the globe itself will be visible.

Where moments ago the somber beige of sand and featureless gray sea and sky stretched, the sun springs like a buoy that has been submerged and released from below. It lifts steadily until it becomes a complete disc. The sun's path is not quite to the edge of the water as a wave rises from the one dark band remaining without light. Its crest colors, climbs, and drops with the color absorbed and clinging to its push up the sand.

In the thin film of rose at my feet lies the transparent oval that is all that is left of a small jellyfish, which I retrieve and dip in the surf to free it from a layer of sand; whatever color or tentacles that it had are gone. During my first summer by the sea I found many of the firm ovals on night walks, their presence made known when I stepped inadvertently upon them, and they glowed with a moon-like bioluminescence. In the Sound and marsh I have seen jellyfish anchored in the lull of wind and in a tide that has slackened, but whether those were the same kind I do not know.

On the underside of the little jelly that fits so easily in my palm is a border of parallel lines. Not immediately noticeable because of this design showing clearly through the colorless gelatinous body is a much shallower and wider fluting on the top, similar to the undulation of waves. The sun coalesces and spreads its soft fire in my hand, touching down in the transparent body with a beauty as spontaneous as that with which its path has come suddenly to the sea beyond it. I hold a silent world of waves illumined by sun. Vastness for one brief and startling moment contained.

Along my early morning walks, much of the life at my feet has been without the barest framework of reference that would reveal what belongs only to the sea and what is found in disparate worlds; much of the aerial life around me has been encountered too briefly to name. One morning during a spring visit, I walked along the beach before dawn. Only pencil-thin lines like those laid the day before were on the smooth slopes of wave-washed sand, the air warm, the crescent moon shining. I had thought the calmed sea had emptied itself of the hordes that had been left in windrows during the week before as the result of a storm.

The first tinge of dawn lightened the sand near the water and revealed the shape of a sea cucumber, a limbless relative of the starfish, stranded on the glistening slope. Rough to the touch, without obvious features that indicated it was an animate being, the animal's faint wick of life changed the inflated body very gradually. When I left the edge of the water and ascended a terrace strewn with swaths of broken shells, more of the amorphous creatures appeared. Full windrows on the terrace belied the calm sea. Although much of the life that the storm had dislodged from the ocean bottom had previously been brought in and had vanished, the local play of currents over a favorable configuration inshore continued to carry in, as a delayed aftermath of the storm, what was not usually found on a beach or found in profusion.

Because of narrow terraces, which days of high and rapid waves had kept frozen in their winter profile, I had taken less time than usual to reach the row of old, weathered houses south of the hotel. They were the first in the area built along the sea. Spare porches girded them, and striped shutters flanked their windows. Although the Sound to the west of a system of great dunes had initially drawn families together to summer, the sea eventually began attracting a few to build near it. They had followed some interior reverberation, as visitors do today, some enigmatic dualism taking them eastward. Designed with a simplicity that imparted a classic elegance to the unpainted exteriors steeped with salt and years of exposure, the houses had aged with an easy grace.

Their beaches collected mounds of weed and sea life more varied than on others that I walked and came to know during my visits to the coast. As dawn began to break, I saw that plant and animal fragments, whole organisms, and those nearly whole and unrecognizable because of missing structures had been given up after the sea's brief respite and were laid in profusion at the sea's edge. Round objects of the same size that I mistook for wave-worked shards of shells began to repeat themselves too often to be random fractures. When I discovered several lying together and stopped to examine them, I was surprised to find that their smooth surface had a rubbery, synthetic feel;

all bore a depression or "well" in the identical place on one side. The riddle of the origin of the objects was solved farther along the terrace by my finding one with a small white shell attached, protecting a compact mass of organs sunk in the well. It was an Ear Shell, coiled and flattened upon what was an oversized white "foot."

A stroller is usually alone on the beach before sunrise. If an enveloping mist has come with the night, it takes land and sea and merges them; all boundaries are lost in a deceptive web of moisture that draws everything into amorphous intimacy. The cricket that is hiding in the dune grass, the egg of the Sulphur Butterfly, the Ghost Crab of the beach, the gull, are veined with the same pale blood. When the dawn is clear, another distinct world is poised to open, and a page—freshly filled with windrows or left empty—is for only one pair of eyes. Solitude sharpens the unseen presence of the sun lying below the darkened horizon, sentient, ready to end the night with a quiet curve of fire. Terraces are still cool and braided with shadows, and the skin waits for the light and warmth it will share with every grain of sand the sun touches.

I did not realize the stretch of beach by the houses was being walked by anyone else. A woman, tall but not angular, came over with a deliberateness that monitored her steps and paced them with the measured way in which she asked what I was doing. Engrossed as I was with some object that was strange to me, her question seemed surprisingly direct. Without introduction, unable to respond with the same ease with which she continued without my answering to speak of things she wished me to hear, I met Nellie Myrtle. She spoke about the dunes, about the tire tracks, and about changes in the beach she believed at first that I recognized. Several times she stopped and asked if I understood what she was saying.

The walk I thought would be solitary brought me into the company of a resident who long ago had absorbed the ways of the sea. With an attention to details and a continuous pattern of searching this beach during every season of the year, she knew well the moods of the coast and the varied life brought in by waves. Her patient, careful eye and her attentive ear were meant only for the sea and its drift lines, and not for the questions I attempted to ask of her. From what had seemed to me an empty yard of dry sand I would have not thought to investigate, she drew a few small, sturdy Coquina shells that by chance were just below the surface.

In her hand she carried a crab I had never seen before. Whenever she approached the edge of the waves, she remained aware of the sudden, random sweeps of foam and moved with them automatically, however riveted she was to something she had just discovered, as if a part of her was merged with the rhythm. I have visited her from time to time to learn from her wisdom and to be sure I understand her. A collection of things she has found on her beach is placed on simple board shelves within her house, which lies just across the road. Shells that are unusual because their patterns of growth are visible and easily followed or because they bear the work of other animals are neatly arranged along the walls of an uncluttered room. The collection is not so much the forum for the rare and perfect shell as it is for what a shell tells of the animal within.

During the Ash Wednesday Storm of 1962, the surge passed easily over the beach road because it had been built level with the sea. The higher bypass behind her house had forced water to rise instead of allowing it to dissipate naturally toward the Sound. The elevated road had functioned like a dam and forced the sea northward, flooding the buildings lying between the roads. Watermarks several feet above the sitting room floor are a dull wainscoting at the level where the thwarted surge had been contained. Inlets have come and have gone in her lifetime, great hills of sand unique to the area have run and flattened until only Jockey's Ridge remains; the old hotels and cottages that were on the Sound where she played are no more because of the shifting island.

She keeps no grievance with the sea, understanding as she does that it is not essentially hostile. It is the sea where she wishes to be buried at dawn, her ashes scattered from a boat launched from her beach. I can imagine it no other way.

On the spring morning of lavish windrows and my meeting Nellie Myrtle, light had come to the sea as a slow flood of color held firm and cool under the rise of a wave near shore. Gradually, each new pool surfaced within the troughs of low waves as gulls shed their somber outlines and swarmed full of light over the sun's path on the water, leaving the terraces unscavenged to fish the sea. Pairs of long Razor Clams as thin as glass were scattered among the windrows. Miraculously, they were still attached by their hinge and still jacketed in a papery skin of greenish brown.

These shells might have been disturbed days earlier during the storm. The shorter, thicker, and more durable shell of similar proportions, the Stout Tagelus, commonly found along the shores of the Sound, would seem to be the likely one for an existence within the restless turmoil of the surf. Yet, the fragile Razor Clam not the Tagelus is the dweller of the sea.

Perfect oval shells that I had found intermingled in weed lines close to the water I found again, as flawless and undamaged, higher on a terrace. They were lying on open sand and perhaps had been brought in by storm waves and left after the water receded. These miniatures of some of the abundant clams of the region were no larger than the nail of my little finger. Difficult as it was for me to lift the thin shells individually from the sand without breaking their delicate edges, some structural immunity as surprising as that of the Razor Clams had kept them whole in the grip of waves.

Numerous tiny fish, intact, were cast up on the sand. Flies darted at my feet, focusing my attention on organisms I otherwise would have missed—the dwellers of the sea, or of the marsh, easily overlooked when trails of debris between windrows were too widely scattered. Deep in the windrows were tiny starfish. Varied forms of life whose final stages I might have recognized were perhaps in a metamorphosis that made them just another mystery in the sand.

In the vicinity of the old hotel, I have wondered about mysteries in windrows and those in the sky that disappear too quickly. As I walked back to the hotel on that same morning of placid waves and full windrows, a group of gulls continually moved ahead of me when I came too close, resetting each time after flying up and leaving traces in the sand where they had been—a few feathers moving in a light wind, the tracks of webbed feet among thin white stains. The sun's fire glowed in the midst of the group's tight, restless shifting of pale grays, dark browns, and white that screened two birds with bills the color of the rising sun.

The glimpse was not enough to identify them. The gulls rose, hiding the pair as all reversed direction, wheeled low over the water, and moved out of sight down the beach.

Had I heard the cry of the two birds that seemed to have fed upon the sun, their voices would have made identification instant. Whenever the high, clear call of the Royal Tern comes to my ear, I think of the bird's intricate balancing of disparate worlds. This tern has a single note dipping in pitch at the end that resembles a hinge working and swinging open a door. If I am by the sea, or if I am at soundside, this sudden familiar call gathers the rich ambience of a barrier island and heightens, in a way nothing else does, an aura of intermittency.

On any clear summer day the lifting of the dark veil over the shoredunes precedes the slow, westward lean of shadows that will hourly be pulled farther and fainter into morning; they will close and deepen at noon before turning east and battening with the glow of the dying sun. A distant focus for light usually seen far out over a morning sea and again in late afternoon when feeding activity intensifies is apt to be a Royal Tern. The tern is as large as some gulls, but no gull wears a black cap and crest or has a reddish orange bill.

The wings, back, tail and breast are a stark white and are highlighted in a cloudless sky with particular brilliance. Peculiar plays of light can sometimes illumine the white area of gulls, giving those birds a reflective brightness that catches the eye. In summer, white feathers replace part of the black cap on the forehead of the Royal Tern, increasing the area of plumage that focuses the morning and evening sun.

Royal Terns gather in the largest numbers of any colonial nester. They nest north to Maryland and as far south along the Gulf of Mexico as Louisiana. Studies of large colonies indicate that a high percentage of nests contain a single egg, and it is usual for parents to hatch but one chick. Royal Terns are resourceful foragers of current edges as much as 50 miles from shore. They also exploit a wide range of inland sites, such as tidal creeks, inlets, and tidal coves of Sound and marsh. The reduced clutch size suggest the possibility of selective foraging techniques, which would be unlikely in the case of birds with more than one chick to raise.

A large, crested tern much less common but similar to the Royal once nested in large numbers some miles south of Nags Head. Colonies of Caspian Terns are typically found farther inland in Canada and around the Great Lakes, but sporadic and localized nesting records make summer sightings a possibility. In spring and fall Caspians are regular but uncommon migrants, often found in the company of Royals. When silent, the two terns are easily mistaken for one another. Despite the superficial similarity, a vast behavioral gulf exists between them. Caspian Terns usually raise several young. The offspring do not look like Royal young and they respond differently to danger. Young Caspians are grayish-white and indistinctly mottled with pale gray on the back and rump. The legs are always black, the bill much lighter. Unlike the darker camouflage of the young of most other colonial nesters, the disguise makes them virtually invisible on the sand they use to hide upon to escape detection.

Caspian adults exhibit only a moderate skill in foraging widely divergent habitats, and should a visitor watch Royal Terns with some regularity, learning the greater speed of their wingbeat, the tendency of Royals to fly higher over the water than Caspians do, to plunge dive into waves from this height, and to generally maneuver with the greater agility that their swallow-like tails and slimmer bodies permit, an intuitive sense of difference can emerge.

Technically, the Caspian is the larger bird. Even for an eye practiced in judging differences in size between visually similar species, the length of the Royal, 45 to 53 centimeters, and that of the Caspian, 48 to 58 centimeters, form a very close overlap. The nearly identical silhouette of beak and head extended to the shape of a rectangle by a crest, the occasional gradation of bill color to orange-red in the Caspian (normally a deep coral red) and the same general coloration of the body make the two terns difficult to separate. In instances where the birds are in flight, the Caspian's shorter tail with its abbreviated notch rather than a deep swallow-like fork and its darker wing primaries are characteristics that are not always decided easily, difficult as they often are to confirm before the birds are out of sight.

At any season, the call of the Caspian hints at the different course taken in its life history. A croak without ornament is the vocal signature of the Caspian.

Royal chicks do not always hatch disguised in subdued colors, as Caspian chicks do. Individual Royal chicks may bear bold orange on one or more legs, on the whole or just part of the bill, or solidly on both bill and legs. The instinct to crouch and remain immobile at the nest site soon gives way to a very different behavior to deter predators, which has no need of camouflage.

In response to danger in large colonies, streams of little bodies flow from the edge of the colony, merge, and move to open ground as a "creche," the French word for a manger. Splashes of bright fire on the feet and bill of some Royal chicks then flicker within this solid mass of birds in one of the most magical harmonies on a barrier island. Where space permits, a thousand young may slide over the sand as smoothly as the sun-sparked froth of a wave. When the molecular pull among the young birds relaxes, the group disbands with the speed of a spent wave vanishing from the sand. Visual markers of orange, black, or buff, combined in a variety of patterns, help parents to locate their own chick for feeding among groups of other hungry birds within a colony. Parents probably recognize the begging cries of their offspring, and what is both heard as well as seen accurately identifies an individual chick.

Whether or not the mastery by the adult Royal Tern of worlds so dissimilar has been fashioned by the needs of a creche is still unknown. The marked difference between the history and habits of the Caspian Tern and the Royal makes the connection attractive. Selectively hunting the inland waters of the Sound and marsh as well as distant feeding grounds at sea to obtain food of the optimum size and proportions would seem to benefit a creching species. Large fish and bulky catches are difficult for young Royals to swallow, which leads to pirating by gulls and other Royals. The pace of a creche is so swift and the creche so large that small chicks are often injured by other young, or are singled out by predatory gulls if a colony is disturbed. Selective foraging by parent terns may provide a way to maintain an even rate of growth that enables their chick to keep up with its creche mates.

The phenomenon of a nursery of young is shared by another colonial nester, the Sandwich Tern. But this tern has only moderate skills, its colonies are not as large, and studies of colonies in Scotland show it does not always creche. The Sandwich Tern is less like the Royal than the Caspian; it is smaller, and the bill is black with a clear, lemon-yellow tip. Like Caspians, Sandwich Terns prefer the company of Royals, but it is the Royal Tern that a visitor is most likely to see, and is the bird I would call the sun bird for the color of sun shining in a manger of young as effervescent as foam.

After the whole colony leaves, the immature Royal is reluctant to fend for itself. It will maintain the parental tie even after migrating to wintering grounds that lie farther south. Before Royal Terns migrate in the fall, it is often a young bird's persistent begging cries to be fed, coming from the beach, that attract my attention and not the familiar call given on the wing, which is so simple a sound for a bird so complex. By fall, a juvenile is as large as an adult. Despite the attempts of the parent to draw it out to sea, the younger bird continues the habit of approaching in a crouched position with open beak and loud cries.

Bright orange flesh has turned black and will remain black if the juvenile was hatched with legs the color of sun. The orange that earlier might have blazed on the bill has faded, but it will eventually return tinged with red and will stay with the bird. The internal valency that once propelled it among other small birds in a sheet of motion is just a memory, perhaps not even that.

8

*T*HIS SPRING DAY I WONDER IF THE fragments of Sargassum, rootless and buoyant like a Red Phalarope at sea and carried to the beach of the old hotel, have felt the slight pressure of lobed feet.

I have never seen the Red Phalarope, evolved from some shorebird group that was terrestrial in habit. It lives during the winter months on the sea. A thick protective surface of feathers overlays a denser undercoat of down; down-trapped air accentuates a buoyancy light as a cork on the water.

This bird in migration is known to feed upon life fastened to floating mats of Sargassum plants, running along fronds in search of food hordes hidden in their crannies, much as it is said to hurry over its nesting grounds in search of insects. Land-felled this morning by the vagaries of the currents, the sprigs that were once part of the golden-brown Sargasso Sea have carried tiny invertebrate riders along the Gulf Stream, over the edge of the continental shelf, and across inshore waters to land. The subtler treasure of a phalarope's foot having run upon the fragments of a floating sea of plants is only for the imagination.

The illogical slenderness of this sandpiper found in storm-roughened stretches of the Atlantic far from shore has made an early sighting recorded aboard ship one of amazement and disbelief, for the sea is described as pushed to hills of spray by a half gale while floating Red Phalaropes. The same record notes that the birds were paddling placidly in the calmer valleys while keeping a position constantly upwind.

"Sanderling of the sea" is the name that describes

the general appearance of a Red Phalarope during the months it spends at sea. After a short nesting season, it loses its deep rust underparts; the entire length of the yellow, black-tipped bill shaped like a Sanderling's darkens to the color of a Sanderling's; the short wing with a white stripe is a Sanderling's wing. Except for a dark line at the eye, the Red Phalarope's winter plumage of gray on the back and white on the belly is suggestive of that of a Sanderling.

Without the fully webbed feet, short legs, and long narrow wings so characteristic of other pelagics that live at sea for most of the year, without a strong fishing bill such as the those of petrels and shearwaters, without the plump and chunky shape of the little Dovekie, the sanderling of the sea is able to become a spinner on the waves. It rapidly gyrates in tight circles by paddling its lobed feet, a motion that brings to the surface multitudes of tiny creatures upon which to feed.

Phalaropes are not rare. Their presence within coastal marshes and upon nearshore water, however, is rare. Because it is the most maritime of the three kinds of phalropes, the Red Phalarope is the least likely to be encountered within the marshes or in the surf zone. Only a brief interval exists between its migration to land in spring and its return to open water in fall, which can be 500 miles from land. As early as the first weeks of June, on nesting grounds located farther north than any shorebird, the exodus by the female back to sea begins. The loss of the fuller-colored plumage of the female in her neck and under-parts, the erasing of her white face oval, her black crown, her dark back feathers bordered by buff, and the same loss in the male starts by July. Despite the brevity of the land period in the life of the Red Pha-larope, accounts of this bird tend to focus upon the bizarre reversal of the normal sexual dimorphism in birds, the female Red Phalarope is larger and brighter than the male, and upon an equally bizarre reversal of courtship roles. Well studied and commonly cited, these traits attract the attention of the curious and eclipse the greater part of the lives of the two birds.

During the time that they spend on the sea, they are transformed to identical plumages of gray and white for spinning on vast and trackless waves. Of the waves of spring and fall that have broken just beyond the door of my room at the old hotel, which of them have floated the sanderling of the sea, able to tune itself to open ocean as closely as its namesake does to a wave's whisper on shore?

What silent enclosed shallow of the marsh has held it buoyed high at the shoulder and whirled by a swimmer's foot? The fortunate come upon the Red Phalarope in the marsh, should the bird be brought there during migration by some accident of wind. Rarely, concentrations of food lure flocks to inshore waters. The revolutions performed to pull tiny animals to the surface are not necessary for a swift and sure identification. It is enough to merely find a phalarope sitting in the water, something typical shorebirds do not do, for their feet are not adapted to swimming. Of the three different species of phalaropes found along the east coast, the Red Phalarope has the widest margins of flattened skin around its three front toes, and a partial webbing between the toes that is proportionately greatest. Even the vestige of a rear fourth toe has a wide lobe. These adaptations are useful for its life on the open sea.

By creeks, ponds, and rivers of circumpolar tundra

the male does the nest building, the incubation of eggs, and provides for the young, a behavioral pattern found in other shorebirds. He departs by early August, often before the young are fully fledged and able to fly. Juveniles are the last to leave the tundra. Stirred by an instinct that sharpens, grows insistent in their untried wings, they follow an unmarked path to sea. Red Phalaropes have been seen off the northern tip of Labrador in the latter part of July. Generally, the Bay of Fundy in August is the staging area where they gather in abundance; invertebrates in the muds and sandflats are numerous then.

From late August to late September Red Phalaropes occur offshore at Cape Cod. They feed near current edges or "fronts" where zooplankton are trapped and held naturally, such as the front by the eastern edge of the Labrador Current. They feed, too, wherever jellyfish, small fish, and crustaceans are in abundance. Or they feed in association with the great whales, partaking of the same concentrations of tiny animals called "krill," and, for the brief time in which the whales surface for air, alighting on their wide backs to snatch crustaceans from the rough skin.

It was thought that the winter range of the Red Phalarope was confined to areas in the Southern Hemisphere where our summer follows the sun, the same places east coast migrant shorebirds go that do not winter in the milder latitudes of the Northern Hemisphere. But some Red Phalaropes cross the Atlantic and winter off the west coast of Africa. Some cross the Isthmus of Panama to the Pacific Ocean and continue on as far south as Patagonia and New Zealand, arriving at a latitude of thirty degrees by mid-August. The Atlantic Ocean is the favored wintering ground wherever concentrations of food are steady. The obscurity and elusiveness of this bird at sea, whose range went undetected for so long and later was only poorly understood, kept the northern part of its Atlantic wintering grounds undiscovered until recently.

Red Phalaropes are now known to gather in flocks of over a thousand birds on a bleak winter sea just forty miles off Hatteras. The flocks are composed of female birds. There they go mostly unseen, their high brief call unheard except by the fishermen of these barrier islands, who go to the edge of the continental shelf and beyond in winter. Before the noise of a vessel alerts the wary flocks and sends them flying, they are likely to be mistaken for pelagics commonly seen during the bitterest months of the year. Phalaropes have the habit of flying low over the water, and are sometimes confused with migrating Sanderlings. If undisturbed, the females float like ghosts on the rising crests and in the sinking hollows of waves, not many miles from the Hatteras shore that flattens peaks and smooths troughs to foam propelled over sand.

Through the salt-rimed window of my room, the chance of finding an occasional line of migrating ducks known as Scoters keeps my eyes seaward. The in-between-times of spring loosen and then harden the air from day to day. The sky seems tightly closed, gray and colorless from horizon to zenith. An iridescence fires upon distant waves, without the slightest trace of sun through what seems a solid shield of clouds. Far away the sun finds a way to the sea, and for a few seconds molten pearl, globular and cool, congeals on the cold water, shimmers without heat or flame, and is gone from the wave tops.

With the sea just beyond my door, I hear the sound of waves gathering, and falling, and gathering again. The pattern of the sound means the waves are low, very long and narrow, formed quite close to shore because of a deep runnel and curled into a smooth barrel to break in one sustained stroke along the beach face; the distinct pause before the sharp explosion of water tells me this. With the swell of the incoming tide, the impact of waves coming down the beach in early spring has the sound of volleys from cannons. In late spring, with a wider more regular march over a smoother berm, the advance of scalloping foam from a wave's edge becomes an audible part of the pattern as the wave sweeps landward then seaward.

The sea can be flat on any summer day, almost as smooth as glass, its waves no higher than the width of a hand, its sounds subdued. In fall, as if taking respite from preparing a change of countenance to higher winter waves and greater swells, the sea grows quiescent if a period of warm weather comes without wind. I can hear then the shadows of sound in those fall waves reminiscent of summer. I can imagine the little birds spinning if I listen carefully. A varied world

of waves is enlivened by the thought of the Red Phalarope leaving its tundra life behind in a plumage that
takes on the gray and the white of the sea.

9

*M*Y AFTERNOON WALK TO THE PIER
gives me the chance discovery of a Common Loon, an
adult bird still in its dusky winter plumage. It rests
on the beach slope just beyond the towering and massive supports of the pier. When I come too near, the
bird slides across the sand as if injured, propelling
itself on its breast with thrusts of its powerful feet and
its wings. It reaches the wash from a wave splay-footed
and with wings outstretched and plumage so dishev-

eled I don't expect it to actually enter the icy water. The contorted, angular body changes at the first touch of sea, as if sea and not solid bone reshapes each limb to draw sinew and flesh together.

Metamorphosed in the surrounding foam, the loon floats easily over the shallows and then moves to deeper water. Steeped in the motion of an approaching wave, the diver rises under the arc of narrow wings as if to fly. Slowly ascending into the stretch, body composed and growing taut, it loosens each feather from another as the wave lifts. With a blending of movements that is hypnotic in its grace, the arc of the wings and the climbing swell become one motion and remain joined as the wave buoys the diver aloft.

With an effortless and careless beauty that seems endless in duration, yet balanced only for the time given to the drawing of a single breath, the loon spreads open its shield of plumage at the height of the wave, and refastens the shield while its wings are motionless.

The dance that ensures its armor of feathers is able to repel the sharp cold of the transfiguring sea fills my eye with all that it can hold. The wave passes,

156

the loon dips into a trough, slowly refolding the wings as it descends. How has this mirror of a wave, by some magical feat of alchemy, learned the skills by which it makes the world of the sea its own?

When late fall comes, the Common Loon has already left the waters of northern timberland and has returned to the coast. The supple and rich choreography of the dance that is basically a "comfort movement" allows a loon to plunge beneath the waves of the sea, traveling deep into this pathless world. Of the bird that is often called the Great Northern Diver, little is known in its other less familiar existence. Studies have tended until recently to concentrate only on part of the life history of the Common Loon, that is, the documentation of nesting along the shores of northern lakes and the specific rituals associated with pairing and defending a well-defined nesting territory.

What little is known of its life on the sea indicates the loon's behavior changes markedly on wintering areas off the coast of southern states, in synchronism perhaps to the sea's wealthier and more diverse sources of food that are seasonally abundant and to periodic tidal ebbs, which have no equivalent on a

woodland lake. A time of disbanding comes for the sexes, and individual birds will defend feeding areas. Although migration from northern lakes occurs in groups, sometimes large ones, once loons reach the coast it is usual to see individual birds fishing alone, the bond that paired the sexes for nesting temporarily abandoned.

Old rhythms cease, the duties of nesting, of rearing young within well-known boundaries. The equinox unfastens the nuptial crowns of purplish or of greenish iridescence on the sleek heads of both sexes; it unbinds their white necklaces from their throats, and eclipses in their eyes the deep red irises. The eyes become dark as the shadow in a wave. The squares of silver on the webs of the feathers of their back and shoulders are lost, as well as the ebony surrounding them; lost are the small droplets of silver on the coverts and on the sides of their bodies, and their background of ebony.

Among waterbirds, the diver is rare in that it waits until it reaches the coast before shedding its large wing feathers or primaries for the yearly molt and thereby melds a flightless body to the waves. No longer white-breasted, no longer ebony-backed with silver dappling that reflects upon the flat surface of a lake and mimics the rippling of a breeze, cloaked instead with brown feathers tipped by pale gray, the loon becomes the sea's.

It dances with greater deliberateness than on a lake, the stages slower and more protracted for descending into the bitter cold of a winter sea. I look for the dance each time I come to the sea. A spotting scope increases the chances of finding a Common Loon on the waves and brings the graceful, fluid motions of the dance closer. An optical illusion occurs if an observer keeps both eyes open several inches from the glass. The background of unmagnified sea that the gaze registers as a natural response will at first overpower the image of the loon that is already in the lens. With a sudden, involuntary shift of the observer's view, the mass of sea gives way to the dancer enlarged in a circle of light. Suspended above water in an ethereal floating existence, magnified to both eyes, the loon is alone on a sea of light, alone in the imagination, and, for a moment, immortal in the world of flux.

When a loon is swimming far out among the

waves, the bird is difficult to identify. The immature Great Black-backed Gull can be mistaken for a Common Loon because of its large size, even the black and white adult gull in light too poor for good resolution. An identification made more from the wish to discover a loon than the reality often has to be later revised. With practice, it gradually becomes apparent that any gull sitting alone on the water is very different from any diver. A gull's body is shaped like a stubby basket with an upright handle at each end. Familiarity with the outline, not the plumage, eliminates the bird. Seldom will a gull submerge any part of its plumage. It is not surprising that the comfort movement such as that which true diving birds perform to open and resettle their feathers while on the water is absent in gulls.

The large grebes, the merganzers, the cormorants, and the sea ducks and diving ducks found here have a response to the rigors of diving comparable to a loon's, but those movements are generally more truncated and are very brief, without the supple, fluid grace of the loon's dance.

To find loons on the sea, a familiarity with other

divers has to be sharpened; how those divers carry themselves; how they compare in size. When I decided to attempt to know the confusing birds found on the sea and, with the right opportunities, began to sort them, the task was very difficult. The increased magnification of a spotting scope more than once revealed that a loon I had been watching was not a loon at all.

A Common Loon is the largest bird seen on the ocean, except for the pelican. The bird most likely to be confused with a loon is a cormorant. Its size approaches that of a loon, and the bird is dark-backed like a loon, is apt to swim alone, and will disappear under the surface. The light area that is usually present at the base of a cormorant's bill does not always hold true. However, the bill is thinner than a loon's and is hooked. The solid dark plumage of the adult cormorant is absent in juveniles, which have the whitish wash on the underparts that is typical of juvenile loons and of adult loons in winter plumage.

I have not seen cormorants peer underwater as loons sometimes do between periods of diving—the fishing device by which loons fall back while slightly elevated and, uncurling the neck, slide the head below the surface in line with a horizontal body. Loons half-swim, half-dive effectively in this way because they are able to stay lower in the water. Cormorants consistently float higher. In the Sound, they are the birds found sitting on a post or on a buoy with their wings spread to dry.

Pelicans, gulls, cormorants, grebes, merganzers, and sea ducks sit higher on the waves; loons merge with waves. Yet, should a loon take flight from the sea, the typical foot patter will commonly be absent that is characteristic of its launching itself from the calm surface of a lake, where its body is too heavy and wings too narrow for lift without a considerable effort by its feet to clear the water. Flight from the waves is stronger, surer; the play of wind over the expanse of sea powers a more rapid elevation here whenever the bird is positioned downwind.

If a loon takes flight more or less opposite the observer, the bird will show a humped back and a dipping of the neck and legs and will confirm a suspicion that what has resisted a sure identification while on the water is, in fact, a loon. Grebes have the slow flight and the humped profile of a loon but are half its size. Merganzers in flight display a more or less even alignment of back, neck, and legs; their wing-beat, like that of sea ducks, is more rapid than a loon's.

If a loon floats close to shore and waves are high, its presence is apt to go undetected. The loon is secretive here, aware of anyone nearby. It is able to stay out of view by sinking until only its neck and head are showing, and the bird will choose the trough of a wave in which to surface after diving, using the rise of the waves nearest shore to further obscure it. Or it will dive and will put new distance between itself and a stroller who has caught sight of it and tries for a better view. In this situation I have waited for the loon to disappear and have used the period it was underwater to get nearer.

During spring and fall migrations, the opportunity to find different kinds of divers close to shore and to find them together is greater than usual. In the vicinity of the hotel, I have found that loons establish the pattern of traversing north or south, working their way on the surface between dives during the morning and late afternoon. Regular sightings of this kind are valuable in learning to recognize the bird. The chance passing of a known bird near the subject thought to be a loon will give a useful reference for scale, helping an identification or eliminating an inaccurate one.

Common Loons and Red-throated Loons as juveniles or as adults in winter plumage are very similar. In these plumages and at a distance, distinguishing the two species can be troublesome. The size difference cited in standard descriptions is not always reliable. Immature Common Loons that for some reason have not grown as rapidly as they should during the rearing stage and those immatures and adults that have been subject to stress in the course of migration or in the period following migration will be smaller than normal. They can then fall into the size range of the Red-throat. However, the bill of the Common Loon is always larger, sturdier, and does not show in profile a straight line at the top. The Red-throated Loon possesses a straight line at the top of its bill, which gives the whole bill an upturned look, a "curve" that is actually formed by the slight lift toward the end of the lower mandible.

The winnowed, trailing cry of the Common

Loon, begun as vaporously as it ends, has been heard far out to sea where observers have discovered rafts of these birds. Although birds forage as individuals during the day, juveniles form rafts, and the sexes are thought to gather in separate rafts at the end of the day. A Common Loon is normally silent for its time on the sea. On both wintering and nesting grounds the Red-throat is quiet, clucking rather than calling, and is without the repertoire of notes the Common Loon commands.

Mistaken ideas of size and shape, and tricks of light or lack of light chasten confidence from season to season, but the time does come when something electric is in the form that is alone for miles and miles of sea. And it is a Common Loon.

Before the hotel opens in early spring or when it is active with guests in summer or after it closes in fall, I nearly always see a Common Loon that is far out on the waves and swimming alone. From the vantage point of the wide veranda by my room, I am able to settle in a chair and watch for the bird that has become the talisman of the building. Each time it is such a tenuous strand of luck that lets me catch sight of the diver between cresting waves, so easily is its presence there unnoticed.

Although Common Loons are not frequently seen in summer, individuals are sometimes attracted to the vicinity of the fishing pier. The danger exists of their becoming entrapped in broken fishing lines, often still bearing a rig of three hooks. One morning, a young Common Loon came ashore entangled in such a rig that it had picked up while swimming or diving. From the second floor veranda where I had been sitting and watching the distant sea, I noticed people gathered by the low pillars of the conduit close to the hotel, with the loon in the crowd's midst. By the time I reached it, some of the bystanders had already managed to remove the three hooks of the rig: one from a leg, one from a wing, and one from the neck. Still surrounded, the loon was wary and alert, unwilling to move through the crowd. I lifted the loon and brought it to the edge of the water, holding its wings to its sides. When it saw it was close to the waves, a full, vibrant note and then another sounded from so deep within the heavy body that I felt them resonate in my hands pressed over the wings.

The notes had the trailing quality of the call I had heard only while in a northern wood. Inches from the water the loon stretched and flattened in line with its extended neck, eager for the first sweep of foam. It reached for the shallows, anticipating escape. Immediately after I released it, it dived and headed swiftly toward the first line of waves, where it disappeared. Long after I lowered it into the sea the reverberation of those notes was in my hands and the enchantment of holding the diver that dances on the sea in the remembered touch.

On land, a loon must shovel forward with its breast pressed to the earth, pushing with its feet to move itself, and if sudden speed is necessary, driving with stiffened wings over the ground. The main bones of the legs are bound within the body by strong muscles. This forces the feet so far to the rear of the body that a loon is unable to stand erect like a goose or a duck, giving any loon an injured look when out of water. The placement of the leg bones that is ill-adapted for movement on land makes the bird the swift diver and swimmer that it is.

This diver that apportions its life between sea and a woodland lake does not come ashore for the part it spends on the sea, unless weakened. For lengths of time that would drown other diving birds, it survives entrapment in shorenets or in nets plied by fishing boats. During migration the number of loons caught in nets increases. Prolonged exposure to cold because of net-damaged plumage or plumage clotted by oil, and poisoning by the ingestion of oil while it attempts to clean its feathers may bring the diver temporarily to shore. On one of my spring visits, I kept track of an adult Common Loon in black and white plumage that remained near the conduit. It had come ashore at intervals as it moved south of the fishing pier. Although the old hotel would be mostly empty for a while, the doors and windows had been unboarded, the tables and chairs taken out of storage and placed on the verandas in front of the rooms.

The bird might have been the darker of two that

I saw from the veranda a day earlier as they swam beyond the bar; it performed a fragment of the dance then, hardly rising out of the water. A linear patch of ocean had for several days the occasional flash of fish and swarms of circling gulls above, betraying the presence of a large school of Bluefish just beneath the surface. By dusk the loon was resting on the near side of the conduit. From the walkway leading off the steps of the first floor veranda, I could make out the bird's form on the sand. The gleam of the lines of dull yellow lights burning on the ceilings of the verandas and the east-west corridors did not disturb it. In the morning when I settled in a chair outside the door of my room, the loon was still by the pilings of the conduit where it had spent the night. With the sand warmed by sun and the sea relatively calm, the loon started to enter the water; I thought it would dive beneath the low waves to begin feeding. A tongue of foam surged over the beach slope. The loon tucked in its head and drew taut as the water passed over it.

It did not dive. Emerged again it performed an awkward, grotesque dance without cadence or arc. The loon was either not strong enough to push beyond the surf zone into navigable sea or did not wish to, and the unremitting surges of waves spilling forced it back to shore. In a sheet of foamy water where it might have begun a dive the loon endeavored to stand, spreading then flapping its wings to bring the body upright. With considerable effort it extended itself above the shallows, wings beating rapidly to achieve elevation until it fell forward on its breast.

A stroller coming down the beach from the direction of the pier was hidden from the loon's view because of the conduit. Unknowingly she came between the loon and the water's edge. She stopped only a few feet from the bird and walked on. A stroller came from the other direction, walking away from the water and heading toward the pier. The loon saw her and scuttled into the cold, bubbling push from a wave. Another wave crested and dropped, and the loon slipped through its falling weight. It surfaced and dove again before a second wave fell. The loon was unable to gauge its movements to evade the roil of surf as it headed toward the calmer zone that lay beyond the line of breaking waves. Gaining smoother water, it moved northward toward the pier, rising only briefly in a sad travesty of the graceful dance a loon performs on the sea.

Fishermen began to gather, waiting as they did the previous day for the school of feeding Bluefish to move over the bar. Strips of sea continued to flash with darting fish, and huge swarms of gulls hung over patches of sea that rippled with tell-tale signs of carnage. At mid-morning the school veered southward, driving before it layers of Menhaden and other small fish. Once it came over the bar, it was in reach of the lines of the fishermen who had been keeping watch on it. Within a growing mass of fishing poles were playing dogs and children, and onlookers who had not dressed for a sudden stop at the beach. Bluefish whipped in the shallows that shuddered with small fish.

I walked to the pier where the more graduated bottom caused long lazy waves to crest. Bluefish streamed into the clear jade eyes of waves, and hung within the vertical walls as the waves were poised to fall. I came upon the loon in the sand in the midst of fishermen and poles on a crowded beach and was able to see the extent of the clotting and matting of its plumage. Water swept in, washing the loon. It lifted its head and called a full, trailing wail shifted to the minor chord, filled with pathos that only human ears create. It sang its own requiem, reaching into the lives of those who stood listening.

A walk at night along a deserted beach can send from the darkness ahead a black scuttling phantom that by its size and speed makes the encounter a startling one. This swift shadow can be a visitor's first sighting of a loon in the setting of a barrier island. Loons are illusive, too, on northern lakes, diving to avoid a close approach, calling unseen at night with yodels and wails, only to the human ear strange and alien, whose drum accelerates the notes of their cries, raises them in register, and accords them tremors of eerie melancholy.

But much is known, nevertheless, of the inland life of the Common Loon or Great Northern Diver. In northern timberland, ledges of rock, gravel bars, banks of earth and moss and grass, even mats of sedge and of floating muskeg are used as nest sites. At first the nest is just a shallow and unbanked depression of plant fragments close to open water, sometimes near dead water off inlet and outlet streams, or along the shores of the islands of a lake.

The position of a loon's legs makes building a nest in the usual way very difficult. The weight of the body compacts the collection of plant fragments and deepens the nest; the edges are gradually built with what vegetation its bill can reach when the bird is settled. As the nest is banked and the rim grows higher, access to the nest bowl becomes more challenging. The loon is forced to assume a semi-erect stance in approaching the site. The bird's full weight is taken by the length of the free part of the leg and by the toes, rather than by the toes alone.

When a lake is free of ice, the nest is built, two eggs laid, and, a month later, hatched. Within a few hours the chicks are in water, swimming with bill submerged and eyes directed downward to the lake bottom. A week-old chick is able to "fly" underwater at depths of over ten feet using wings and legs to travel a distance of forty to sixty feet. Aerial flight comes in eleven weeks, and after that only the muscular legs with their blade-like tarsus propel the body in diving.

If the moon is the avatar of the sheltered world to the west in this setting of a barrier island, the sun is that of ever-moving sea. On a sea that held a loon, the fierce heat of the sun once budded shining spheres and dropped them to ride along the wave tops, where they remained and continued to ignite as if struck and restruck from a flint. Wild and prodigal, sprung loose from the sun, the spheres collided with one another and veered away, fastened to a wave crest and set apart from the clear blue of the morning sea by an invisible wall. Nothing exhausts itself. Nothing repeats itself in the same way within this world of contingencies. The loon is at home in flux.

Before a loon dives below the waves, air is expelled that is trapped in its heavy down and in the body's pockets under the skin. Watertight feathers are flattened and pressed tight to the skin; skin is bound tighter to the body. While a loon is submerged, it is impervious to the natural toxins produced by diving. Muscles that are red in color breathe as life is withdrawn from the limbs and peripheral interior organs and is channeled to the nervous system, the master of the senses that orders movement. A taut, slimmed shape with neck outstretched and aligned with the drive of blade-like shanks and webbed feet flared on the downstroke and collapsed on the upstroke takes a loon into the abyss.

I wonder if the trailing, willowy cries and unearthly tremolos rooted in the very density of the diver's bones and heard in northern timberland tell of the mystery of that wilderness. Or does a loon sing there remembering its first and longest passage to the wild sunlit sea and its shadows? As a juvenile wearing the dusky plumage of the winter adult but not its darker bill, a Common Loon leaves the lake of its birth in the first fall of its life and does not return the next year. It will not return the following year. A young loon is at sea for three years.

That the wilderness of the sea is the palimpsest for the life of an adult loon brings me to marvel again over fragile things placed in proximity with the immense and random. I discover a juvenile bird one day as it swims on the waves in a swath of light. April sun glitters on water like small fish shoaling at the surface, the myriads of slender crescents flashing and darting erratically as the sun's vibrant life is mirrored

176

across the pulse of waves. The young diver is the only bird upon the expanse of water except for several gulls, which settle only momentarily with wings upraised, held ready for flight should a crest spill too near. At times the bird lowers its sinuous neck and places just its head beneath the water. Every now and then it dives out of sight and the sea is empty.

I do not always notice where the loon has resurfaced; I find it, eventually, upon the rolling waves. On another day, I see what is probably the same bird as a storm comes out of the northeast, darkening the sky and seaming it with jagged lighting. High black clouds pile one upon the other while a gathering wind begins to scuff the wave tops and deepen troughs. Oblivious to the approach of the storm, anchored, I think, by flightless wings to the plunging surface that is at last hit by a drenching downpour, the young loon stays on the sea.

I keep watch from the protection of the wide veranda on the second floor near my room and see the dark form riding long swells smoothed and stippled by rain. On still another day, I find the loon by a splinter of light that the sun has dropped through a narrow rift in the clouds. It is feeding, darting swiftly beneath the water to swim just under the surface. A large wave crests and begins to fold near the loon still shelling under water. Aware of the oncoming push of the falling wave, the loon dives and is gone by the time the bore of water passes. Within a minute or two of returning to the surface it eludes another avalanche as easily, moving gradually into the core of the maelstrom of breaking crests triggered by the inshore bar. The young loon is soon maneuvering in the swirling backflows rushing seaward and in the constant push of waves that angle over the bar and buckle in a punishing turmoil, evading them all as if invulnerable.

At last, while I am looking elsewhere, the diver leaves the surface and flies. I catch sight of it, quite by chance, as it is still climbing strongly into the air and follow its flight with binoculars, tracking the moderately rapid and even beat of the wings. I expect the loon to drop back into the water beyond the end of the pier. But it heads out to open sea. The dark silhouette becomes smaller and smaller; I lose the form in the close graining of distant sunlit crests.

Part Four

10

*T*O THE WEST OF THE HOTEL LIE THE smooth contours of Jockey's Ridge, pale and feature-less under moonlight, and beyond those great slopes of sand, the waters of the Sound and marsh. On a warm night at summer's end, nothing stirs beneath a high vault of clouds and the August moon that illumines a corner of the marsh the tides of the sea do not fully reach. Because generous rains have come and fresh-ened the brackish water, a flowering plant known as milfoil, named by the word meaning "thousand-leaves," has recently burgeoned and floats in mats of finely dissected foliage and emergent fruiting stalks bearing tiny blossoms.

As my boat makes a path around the patches of milfoil, the oars dipping, a tentative frog piping, a Screech Owl spiraling out a silken call are isolated sounds beneath the moon. Some miles away, the Bodie Island Lighthouse sends a steady, circular swing of light into the darkness of other retreats from the amorphous sea. The boat angles slowly toward the nearby island of Black Needlerush. A cloud edge sliding along the curve of sky covers the moon; the marsh and the creatures it shelters are in total darkness, and I continue on to see how deep their slumber is.

Near the rush bank of the island, a sound that is neither the call of any bird I recognize nor of any frog creeps from the darkness. So faint is the chorus I am hardly aware of the moment it has crossed the threshold where it is loud enough to be heard. The high whine of voices swells gradually as the boat moves parallel to the island. Without any sign of its source the chorus of insects blankets the night, a thin wiry fabric that grows more shrill with every dip of the oars. The brush of wings that I expect never comes; yet, the insects generating the hum surround me, and in the cone of light I flash through the darkness are flying bodies.

They are so agile in eluding the sweep of my hand across the beam that I am unable to snare even one as they whirl and pitch in a dervish dance, crowding the narrowest part of the beam and the widest. The number of fliers grows as the boat moves along the rush, and the night that began so quietly has filled with the sharp whine. The boat glides into an opening on the muddy bank where the island is broadest. Leading from the bank toward the interior of the island, the path that I follow on foot soon disappears. The swarms are thickest here, so concentrated among the stems that the pitch of the hum when I crouch close to the ground is deafening.

How is the mind to imagine such an astronomical number of bodies magnetized to a dance that precedes the laying of an astronomical number of eggs?

The insects are adult aquatic midges. They are often confused by sight with male and female mosquitoes and by sound with female mosquitoes, which are able to vibrate chitin parts on their thorax to produce a whine. The sexes of the aquatic midge and those of the mosquito have banded bodies, long legs, and are about the same size. Both sexes of the aquatic midge have a short beak and by this feature resemble most the male mosquito, which does not have the elongated, puncturing mouthtube that the female possesses. Midges have shorter, wider wings than those of mosquitoes. Although the antennae of the male midge are feathery, the antennae of the female midge and both sexes of mosquitoes are not.

In a day or two the midges will eventually settle. They will lay their eggs in gelatinous strips along the bank of the rush island or in the shallow water near it. Other nights to come will cloak a life as opulent, more elusive for the senses to grasp, a life without voice and that goes unseen beneath the water, which will feed upon the eggs of the midges. The labyrinths of marsh are nurseries, cradling countless numbers of tiny fish destined for the reaches of the sea. And they return here from the sea, if the salinity of the marsh and the adjacent Sound approaches that of the ocean, and corridors are open, the journey of these fish made many times, dividing existence between a world with boundaries and a world that is harborless, always in flux.

By morning the moon is low, refulgent, silvery, only faintly golden, its rays falling upon the water as a funnel of light interrupted by swaths of milfoil puckering the surface. Burnishing slowly to pale red, its vigil nearly ended, the moon slips down through the limbs of a cypress standing on shore where the marsh edge eases into mud and spongy grass. The moon goes from branch to branch like a deliberate bird. Still visibly sculptured, it darkens to copper near a band of haze lying on the horizon. The band drains its fire until the moon disappears and then restores the disc, emptied, a colorless circle scarcely more than its own outline.

At the end of a road near some empty buildings and a group of docks, I find a place one morning in a different part of the marshes to watch the moon end its vigil. A bullfrog in a narrow, brackish channel by the road ceases its solitary pumping when the first light of dawn comes. To the edge of the roof of the nearest building a crow flies. The alarm it gives is the call typical of a Common Crow, without the nasal ring that identifies the Fish Crow. Soon after, a Belted Kingfisher clatters like a stick moved swiftly across wooden pickets, and gulls glide out of the mist like specters.

From the wrapping of darkness by a border of Black Needlerush, a large bird lifts into the air, the beat of its wings slow and stately, the beat of a heron. Conjured by some obscure influence where the wind raises gentle ripples, the moon floats in redundant images. Among three golden orbs tiny fish tipple and dart, perhaps young ocean fish, perhaps mummichogs or killifish that live only in the marsh system. Several derelict docks stretch into the water, their farthest ends mere lines of posts. Earlier, gulls slept there.

From the moment that the bullfrog grew silent, ripples of sound continued to move into the formless dawn, rupturing the stillness. Royal Terns and Laughing Gulls maneuver noisily to alight on the diminishing number of dock posts. It is not long before the rows nearest me are filled. The occupants are tested repeatedly and respond with eruptions of warning cries, beaks wide open, heads tilted forward during the challenges. I witness the rapid dislodging of five birds. The first is usurped after its warning cries fail to discourage the intruder. Flying to the second post the rousted bird succeeds in moving the occupant of this post to the next, where, in turn, a third bird is sent off. It goes down the line and sends the bird on that post flying, which performs the identical maneuver at the next post.

Strung from the nearest dock posts, numerous spider webs emerge one by one from the shadows. Light encroaches into the crevices and corners under the boarding of the docks, revealing more webs. A sailboat rocks in its slip, thumped lightly by wavelets. One huge web breaches the angle between the mast and a rope, forming a sail that catches no wind. Not far from the slip, two white birdhouses on slender poles stand surrounded by water. One of the houses has attracted a pair of Kingbirds that returns to its roof to perch after sallies for insects.

Just beyond the Kingbirds, a Least Tern, scarcely larger, flutters on narrow agile wings and calls to a juvenile bird resting on the planking of one of the docks. Without alighting it hovers close with food in its mouth, uttering a series of short cries before it goes off, the food still in its beak. In the protected closures of flat and rippled water, this swallow-shaped bird trying with a morsel of food to tempt a young bird from the planking seems in place, in perfect harmony with the Kingbirds, the meandering creeks, the tide inching up and down island banks, the soft wind eddied in the maze of vegetation.

Although it has the diminutive proportions of what should inhabit only the world of the marsh, wings and body scaled to a life here, the Least Tern also belongs to the sea. There it hurls itself into wave tops when the sea is up, striking them with a force great enough to fling towers of spray into the air, then shaking from its feathers with the most subtle of tremors what droplets still cling to its body.

Of all the birds that make passages, it is the most fragile to hunt with a slender bill the color of the moon both the marsh and the waves and to build a nest where the sound of waves is never stilled. Three images of the moon still lie on the water, and two of the images vanish as the wind shifts in its fortuitous play.

The rim of the last seals itself against any seepage of light into the saffron surface. Behind me the sun is rising. It lays its fire over the marsh water in a tenuous sheet of mauve, and this touches the edge of the moon as a delicate confluence of worlds.

The plane of the earth's orbit around the sun is offset from that of the orbiting moon by five degrees, a variation too small to explain why the moon has ranged over different sections of the sky during the course of my visits to the marshes.

But, the earth is tilted at an angle of 23 ½ degrees as it rounds the sun. When in its cycle of one month's duration the moon crosses north of the plane of the earth's orbit, the ascent is known as "moon high." When it crosses south of the plane, the descent is known as "moon low."

The offset of five degrees between the orbiting

moon and earth's orbit added to the degrees of the earth's tilt moves the moon 28 $\frac{1}{2}$ degrees north of the plane of the earth's orbit and 28 $\frac{1}{2}$ degrees south of it in a single month. This is a total of 57 degrees. It provides for considerable fluctuation and causes the moon to follow a path that gradually takes it close to the zenith or close to the horizon.

At the time of the winter and summer solstices the range of the moon is even further broadened. At winter solstice the arc of the moon dips closest to the horizon—but this moon, the new moon of the calendar is the "full moon" never seen; the complete disc actually seen is the moon low at summer solstice when the sun rides highest, and the moon high at winter solstice when the sun rides lowest. One morning, under uneven tendrils of clouds high in the zenith lay a background of pale blue. Rose crept along the clouds until it encircled them. At quite a different location in the water by the docks that morning, I found the image of the moon and the image of the sun side by side, the fire of the sun subdued and shed like a molted skin, its life elsewhere on the waves of the sea.

Each fall, sometimes with the moon as a beacon,

scores of migrants from the north fly to the marsh before fronts of cold weather. Snow Geese come on glistening white wings tipped with black, spiraling out of the night in late October or early November to land upon brackish ponds.

If rain has been ample enough to fill the mud pans by the roadside, a sudden layer of geese in search of harbor can cover the shallows that were empty the day before. Until the migrants forage along the soft bottom in the interior of the marsh, their black primaries and the stark white feathers of the body and upper wings will remain unmuddied. But the Snow Geese will not stay close to the road for long in fall. Responding to a signal too slight for human perception, the agitation of a single bird perhaps, they will rise and, in shifting angles, imbricate the sky with a geometry of black wing tips. In slow circles the geese will pull farther away after each turning to feed in the deeper parts of the marsh that are more secure, leaving pond and sky empty once more, the autumn air still. Whistling Swans will arrive in the marshes to winter, and I can find them in tableaus on the edges of marsh pools, very long, thin necks bending languidly back when the figures suddenly come alive to slide black bills along the feathers of their folded wings.

Natural breaks against the wind when it blows prevailingly across large areas of open marsh are islands of Black Needlerush, and in the more saline stretches, swaths of a salt-tolerant cordgrass. This cordgrass has its leaves alternately attached along its stem. *Spartina alterniflora* is the dominant cordgrass of tidal wetlands, able to expel excess salt from its roots, as well as from its leaves and stem, and to grow where other marsh grasses cannot. Its decayed leaves are a source of nutrients, making the marsh a fertile birthplace and nursery.

Cordgrass as well as Black Needlerush form out of the way alleys and provide the retreats for secretive inhabitants that do not wish to be seen. The stretchable form of marsh birds like the shy Night Heron, the bitterns, and the rails allows their bodies to elongate and to travel through vegetation without the telltale swaying of stems. Intertwining creeks flow by low hillocks supporting young trees. Oak, elder, and Groundsel grow in thickets near mud banks that at low tide might bear the tracks of a fox, a raccoon, or a shrew, a mammal small enough to fit within the print of a fox. At one of the blind ends of the Sound, an island can open upon a strange collection of drift: a rusty pail, a child's toy, a sandal, a glass bottle brought from the ocean on the tide.

When they grow too near the sea on low barrier islands, Live Oaks are held by the spray of minute salt droplets to prostrate forms trailing in the sand, the contents of new cells emptied by osmotic pressure. On some islands, the bushes of the hardier elder and Groundsel are scaffolds for the bulky nests of colonies of herons and egrets. Salt spray effectively prunes the tops of these island shrubs if they are directly in the path of the wind, stunting the new shoots of seasonal growth and shaping dense nesting platforms firm enough to be used by the heavy birds. On islands only a few feet above the high tide mark, bushes record frequent winter overwash, catching and holding in their branches odd bits of rope, netting, wood and other debris as the sea rolls through to the marsh.

On islands protected by dunes, Wax Myrtle and Bayberry become tall bowers. The leaves of Wax

Myrtle are half as broad as those of Bayberry. The berry of Wax Myrtle is smaller by half, and its seed occupies relatively less space within the berry. The seed of the Bayberry mostly fills the space surrounding it. The leaves of both bear wax cells on their shiny upper surface, as do the grayish-blue berries. In an enclosed shelter of these shrubs, the fragrance from leaves that I pluck evoke candlesticks and lace, an aura of moons and moorings. In the branches the messages of "writing spiders," inscribed with pure white threads at the center of enormous webs, tell of the palpable intimacy lying within the thickets, of the specialness that comes in quiet places.

Summer residents of the marsh like the Mallard, the Wigeon, and the Black Duck belong to the group of ducks that do not cross to the sea. Adept at surface feeding with the body upended and wings held at the sides, they scour the bottom of shallow water with a wide, serrated bill specialized for this method of gathering food. The wings of marsh or "dabbling" ducks have a speculum or iridescent bar on the rear edge, best seen when the wing is extended. As secretive as the shy birds of the marsh when they choose to be, the ducks travel at low tide beneath the drooping mud overhangs of islands, quiet as shadows in these road-ways.

Dabblers nest early in summer, a time when the Arrowleaf Morning Glory blooming faint blue winds tightly about the cattail stem, lifting its flowers higher and higher, and thin blue-bodied Dragonflies sail down paths ahead; when turtle clans gather and the leathery shells of their eggs everywhere lie, split lengthwise, empty; when Fiddler Crabs scuttle on the loose mat of marsh grass, find their mud holes under-neath, and disappear in the sieve, leaving Periwinkle shells still bound to the grass. The rhythm of life is swifter, surer, then, in pace with the Snowy Egret that on black legs and gilded feet dances to the image of itself it forms and rends with jolted irregular steps.

The world of the marsh is the only world of the elusive brown-eyed grebe, which possesses a swim-ming foot of three lobed toes as other grebes do. Its short bill that bears a black ring near the tip gives the little diver the name Pied-billed Grebe. Its relative, the Horned Grebe, does visit the sea. Its eyes have a vivid scarlet iris. I have seen Horned Grebes, effortlessly avoiding the eddies of breaking waves and keeping beyond the reach of foaming surges. Smooth water seemed to always surround them as if their eyes of fire, rather than the relentless working of their lobed feet, continuously formed an oasis as still as a marsh pool.

The marsh is the world of the finely proportioned Avocet with rusty neck and long upturned bill reflecting in still water like painted porcelain.

The marsh, not the sea, is the world of the Solitary Sandpiper. Small moon-white circles spot its slender dark back. Seen only during its migration, not common then, this bird does not seclude itself in the interior of the marsh to elude the visitor who wishes to view it, but may quietly go about its feeding where the marsh begins near a crowded and busy parking lot.

A sense of subsiding rhythms pervades the hours I spend in the more quiet part of the marsh at summer's end. It is a place where flowers of the mallow bloom, some small, some very large, their huge soft, pink petals drooped by their own languor along a walkway by a pond or by paths near flats of water. Here, the Osprey circles wide on unhurried wings, hunting the marsh creeks until it heads for a lone post in the water. Approaching it with wings outspread behind spined, taloned feet with a reversable forward toe, the Osprey thrusts its feet forward toward the post in the same way it seizes a fish. Smaller than an eagle, this white-headed, dark-cheeked bird with distinct bends at the black "wrists" of its wings can forage the sea and plunge below the surface, unlike other raptors. Its careful deceleration before landing on the post top seems prolonged, as if the final motions have been slowed to those in a dream, before the talons touch, then close upon wood.

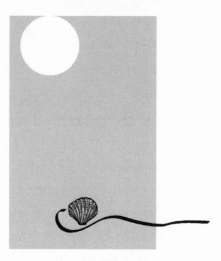

11

*A*T THE QUIET HOUR OF EVENING WHEN the mood is one of ease and sunsets, before clouds turn iron gray and shadows lengthen, the soft globe of the sun sinking in wintry clouds engulfs the sky in mauve, progressively deepening the color and strewing the waterways of the Sound with embered fire, sealing the sheltered side of the hotel in its glow.

This old building that stands by the sea brings me to count it as an improbability, a part of the bifurcated life that endures within reach of waves. Between the wings and the central section, two open corridors on the lower floors join the soundside of the hotel with the oceanside. If the wind is from the east, the strong scent of sea and the roar of surf is tunneled through these accessways that lead from one world to another. The corridors of the first floor have known the swell of hurricane surges through them. Sturdy pilings beneath roughhewn cypress beams support the entire structure and hold it several feet above the sand, permitting high seas to travel unobstructed.

As fully exposed as the hotel is to storms coming off the ocean, the moderate pitch of the large hip roof on the north wing and the south wing offers little resistance to the force of gale winds. The simple and steep roof of the central section is recessed between the two wings. The overhang of all three roofs shields the second floor veranda, which encircles the building.

Viewed from the top of Jockey's Ridge, the hotel is a massive structure compared to the houses and cottages near it on the beach. To the west from the ridge top what appears to be a solid expanse of deep water is really a broad system of shallows, often deceptively corrugated by breezes ruffling it. The coming of fall cools and enriches textures, sharpens contrasts, and breathes into various parts of the Sound startling depths of color, infusing its shallows with inky blue and its channels with black. The sun shifts slightly within its narrow band of sky called the ecliptic, causing light to strike more obliquely. The rush borders of Sound and marsh become dark green, the borders of cordgrass, emerald. Yaupon trees will gradually laden their branches with berries, which throughout the winter months brighten borders with accents of bright red. The fruit of myrtle and Bayberry are strokes of pale blue in places, the hue sometimes dotted with the lacy yellow of goldenrod or with the feathery white pappus of the Groundsel tree.

In the vicinity of a rustic hotel by the Sound, a cottage community of summer visitors was established over a century ago. A pier was built to allow the docking of packet steamers and of boats bringing visitors. For the colony, life was not without contingencies. The initial hotel was destroyed during the Civil War and was later rebuilt. However, it was then buried by the drifting sands of nearby ridges that are no longer in existence. The hotel was again rebuilt but was demolished in a fire. During the early years, others sprang up and thrived by the Sound. None have left any traces of themselves or have any of the houses. A resident, who has written about the first road to open the coast to the automobile, reports that the asphalt for the road was hauled to a mixer near a cottage known as the First Colony Inn. It would seem that the name of an earlier hotel that once stood near the Sound is the same as that of the hotel that stands by the sea.

Structures that were contemporary with the hotels and too close to the dunes were buried; those too close to the shore of the Sound were casualties of erosion and ice. In some instances, houses in imminent danger on a site were moved on rollers to the sea. Relocating from the Sound to the sea was not unusual because buildings were without chimneys and sunken foundations, which made them easier to transport. One of the hotels built on the soundside was moved to the sea. Before it perished, its rooms that had known the more placid tempo of enclosed waters knew for a time the restless cadence of ocean waves. Of the hotels made of wood that were constructed in place by the sea, a rogue wave that suddenly emerged with destructive force during a storm has claimed one; fire, another.

The summer hotel known as the First Colony Inn, built with all the wisdom of resiliency, made symmetrical by corridors and encircling verandas, is the last of its kind. Warped by sea air and faded with age, old photographs of the hotel hang on one of the lobby walls, showing a wide sand beach without terraces stretching into the distance, and a walkway for bathers that is much longer than the one that leads off the stairs from the veranda toward the sea. With the continuing encroachment of the sea, there may come a time when the building will be separated into three parts and the sections placed on wheels to be moved closer to the Sound.

On the top floor each wing has one spacious room that runs the width of the building and gives views of both sea and Sound. The long section between the wings, constructed at the same time as the wings, had rooms that were once very small, merely a series of cubicles that either faced east or faced west. The rooms of this section were made larger, and those on the top floor remodeled to give both east and west views through dormer windows. Winter waves will come and pluck at the cedar shakes and the railings of the first floor veranda on the ocean side. Waves will further damage the walkway laid on sand. In spring when the building is reopened for another season, the cedar shakes will be nailed back, broken railings repaired, the walkway shortened, and old paint freshened with a lime wash. If the building is ever separated and taken down the beach road to a safer site nearer the Sound, the restless sea will again find it, and will restore some interior reverberation, again making it a living part of a barrier island that holds on its ocean side the aura of what is without boundaries, and on its sheltered side what is familiar.

*T*onight, the hotel is empty. The outside louvered doors of the rooms of the first and second floors are tightly fastened, the wooden shutters of the windows closed and secured. For this visit, I have been given a room on the top floor of the central section near the north wing. When I look from my dormer window to the sky beyond the jutting slant of the nearby hip roof, stars and a Hunter Moon are visible over waves gentled by a land wind.

The inside doors connecting all the rooms along the length of this section of the hotel were swung wide a few weeks ago and left open for the winter. Looking down the long line of doors from my own room, I see a distant yellow light gleaming faintly behind the glass panels of the last door at the other end of the building. Shadows flow unchecked. The darkness fills with the murmur of waves slipping from the edge of a calm sea under a wind from the west. Whatever ghosts of the lives once occupying the rooms may be here to await the long winter, they are at rest and listening.

A cold front moves in during the night, altering abruptly the cadence of low waves breaking on sand. A northeast wind wakes me, beating hard under the overhang of the hip roof of the wing and snapping the tattered shades over the open windows. Stars vanish in a rapidly closing sky. Pounding sea under the shift of wind sends clumps of white froth skating across the sand, the messengers of winter. Clouds pale the Hunter Moon, barely visible in the darkness. Shadows whisper, and all the empty rooms fill with the sound of waves rushing over the fetch of sea. My thoughts are of passages and waves not shadows; of the insistent template that makes a human life ambivalent with the need for stability and with the need for innovation. By possessing the strength of fragile things, I am mutable in the world of flux—it is what I absorb from the darkness and the thunder of sea.

I rise before dawn and go out into the blackness.

The mysteries in the sparse windrows on the sand are already written by the high tide that came in the passing night, leaving a broken script as the water receded. Gulls have not yet scavenged the beach; the windrows are still fresh fragments of sentences as I walk among them and endeavor to read what is written there. A few molted crab shells of chitin bearing their runic messages have come tumbling in on the waves and have been left intact, and, too, some of the tiny and fragile clam shells that always make me think of odd juxtapositions. The leading edges of the highest waves to course up the slopes have left them for the next tide.